A Practical Guide to Protein and Peptide Purification for Microsequencing

Second Edition

A Practical Guide to Protein and Peptide Purification for Microsequencing

Second Edition

Edited By

PAUL MATSUDAIRA
*Whitehead Institute for Biomedical Research
and Department of Biology
Massachusetts Institute of Technology
Cambridge, Massachusetts*

ACADEMIC PRESS, INC.
Harcourt Brace & Company
San Diego   New York   Boston   London   Sydney   Tokyo   Toronto

This book is printed on acid-free paper. ∞

Academic Press, Inc.
1250 Sixth Avenue, San Diego, California 92101-4311

United Kingdom Edition published by
Academic Press Limited
24–28 Oval Road, London NW1 7DX

Library of Congress Cataloging-in-Publication Data

A Practical guide to protein and peptide purification for
    microsequencing / edited by Paul Matsudaira. --2nd ed.
        p.   cm.
    Includes bibliographical references.
    ISBN 0-12-480282-6
    1. Proteins--Purification.   2. Peptides--Purification.   3. Amino
acid sequence.   I. Matsudaira, Paul T.
    [DNLM: 1. Amino Acid Sequence. 2. Peptides--isolation &
purification. 3. Proteins--isolation & purification.   QU 55 P8948
1993]
QP551.P648   1993
574.19'245--dc20
DNLM/DLC
for Library of Congress   92-48384
                                                              CIP

PRINTED IN THE UNITED STATES OF AMERICA
93 94 95 96 97 98   E B   9 8 7 6 5 4 3 2 1

Contents

Introduction
Paul Matsudaira

1. Strategies for Obtaining Partial Amino Acid Sequence Data from Small Quantities (<500 pmol) of Pure or Partially Purified Protein
Harry Charbonneau

Contents

2. Enzymatic Digestion of Proteins and HPLC Peptide Isolation
Kathryn L. Stone and Kenneth R. Williams

3. Purification of Proteins and Peptides by SDS–PAGE

Nancy LeGendre, Michael Mansfield, Alan Weiss,
and Paul Matsudaira

Contents

5. Mass Spectrometric Strategies for the Structural Characterization of Proteins
Hubert A. Scoble, James E. Vath, Wen Yu, and Stephen A. Martin

Contributors

Numbers in parentheses indicate the pages on which the authors' contributions begin.

Ruedi Aebersold (101), The Biomedical Research Center, and Department of Biochemistry, University of British Columbia, Vancouver, British Columbia, Canada V6T 1W5

Harry Charbonneau (15), Department of Biochemistry, Purdue University, West Lafayette, Indiana 47906

Nancy LeGendre (71), Millipore Corporation, Bedford, Massachusetts 01730

Michael Mansfield (71), Millipore Corporation, Bedford, Massachusetts 01730

Stephen A. Martin (125), Genetics Institute, Inc., Structural Biochemistry Department, Andover, Massachusetts 01810

Paul Matsudaira (1, 71), Whitehead Institute for Biomedical Research, and Department of Biology, Massachusetts Institute of Technology, Cambridge, Massachusetts 02142

Hubert A. Scoble (125), Genetics Institute, Inc., Structural Biochemistry Department, Andover, Massachusetts 01810

Kathryn L. Stone (43), Howard Hughes Medical Institute, and W. M. Keck Foundation, Biotechnology Resource Laboratory, Boyer Center for Molecular Medicine, Yale University, New Haven, Connecticut 06536

James E. Vath (125), Genetics Institute, Inc., Structural Biochemistry Department, Andover, Massachusetts 01810

Alan Weiss (71), Millipore Corporation, Bedford, Massachusetts 01730

Kenneth R. Williams (43), Howard Hughes Medical Institute, and W. M. Keck Foundation, Biotechnology Resource Laboratory, Boyer Center for Molecular Medicine, Yale University, New Haven, Connecticut 06536

Wen Yu (125), Genetics Institute, Inc., Structural Biochemistry Department, Andover, Massachusetts 01810

Preface

Why a second edition? The first edition focused on methods of obtaining the N-terminal sequence. Since the appearance of the first edition, there has been rapid development of new techniques as well as welcomed modifications of old methods. In a sense, this guide reflects continuous refinements at the cutting edge of protein purification techniques to obtain not only N-terminal sequence information but also internal sequence information. As a consequence, the second edition provides practical answers to a more general question, ''How can I obtain useful sequence information from my protein or peptide?'' rather than the more specific question, ''How can I obtain the N-terminal sequence?'' asked in the first edition. Proteins, unfortunately, often have blocked N termini and, as a result, we spend a large proportion of our time trying to obtain internal sequences.

In this edition we have been careful to present protocols that we have found to be the most useful and use in our own labs. Many changes reflect our better understanding of the techniques, or new developments in instrumentation, reagents, or materials. For instance, blotting onto PVDF occurs through mechanisms that are still not completely understood, but manufacturers have developed new membranes with improved binding capabilities to improve sequencing from blots. Paradoxically, internal sequencing from blotted proteins requires peptides that can be eluted from the same ''high binding capacity'' membranes with equally high efficiency.

We painfully appreciate that proteins with blocked N termini are abundant and we need to develop more practical protocols to obtain internal sequences. Because of additional manipulations required in sample handling, yields of these peptides are reduced. This need has fueled a new effort to define conditions for generating peptides and for recovering the peptides with good yields.

The second reason for this second edition is that there has been a revolution in mass spectrometry. The first edition was published just as this revolution was launched. Three years later, the dust has not yet settled; however, many new important developments cannot be ignored. The mass spectrometer has now become a high sensitivity tool for obtaining masses and protein sequence information. The levels of sensitivity rival those obtained from automated protein sequencing instruments, which have also improved in sensitivity. The mass spectrometer has also developed into a tool for biochemists to characterize peptides. Recent commercialization of laser desorption, plasma desorption, and electrospray ionization techniques has permitted laboratory researchers to operate their own mass spectrometers. This has generated new protocols for sample handling, which have constraints not shared by automated protein sequencing instruments. Chapter 5 on mass spectrometry, has thus been completely rewritten.

The other chapters have been modified, some more than others. Substantially new is the short treatment of what to do with the sequences. This section at the end of Chapter 1 guides you through the initial steps in looking for alignments or designing oligonucleotide probes. While much of the actual manipulation depends on the software package that is installed on your computer, the scheme for analysis is the same. A new section on N-terminal blockage has been included in Chapter 3, following the publication of simple methods to remove blocked N-terminal amino acids from electroblotted proteins. An Appendix that contains the sequences of the most commonly used proteases has been included. It is not unusual to obtain a sequence from an unknown peptide only to learn later that it is the sequence of the protease used to fragment the protein.

As before, we have left out much material but have directed the laboratory researcher to more comprehensive treatments of proteolysis, peptide purification, or sequencing protocols wherever appropriate. This information is useful for the minority of cases in which a more tailored protocol is required by special or unusual properties of the protein that render it difficult to purify or analyze. Our intent is not to provide an exhaustive compendium of methods but rather to provide a set of protocols that one would try initially in one's studies. If our protocols fail, then various encyclopedias of protein purification can be consulted.

I wish to thank the authors of each chapter for enthusiastically participating in this effort once again, the researchers who developed significant new methods that made a second edition necessary, and the people who made suggestions for improving the first edition.

Paul Matsudaira

Preface to the First Edition

This book provides practical answers to the question, "How can I obtain the N-terminal sequence?" To answer this question it is necessary to establish the amount, molecular weight, and purity of the sample, whether the protein is modified (glycosylated or phosphorylated), how the sample was prepared, and the buffer in which the sample is dissolved. One quickly learns that the sequencing is the easy part because this process is usually automated. Most of the work is in purifying the protein or peptide. Unfortunately, the physical–chemical properties of proteins vary widely; unlike nucleic acids there is not a universal cookbook protocol that can be used to purify all proteins. However, there are procedures that can be followed in most cases.

This book is a result of a workshop held at the joint meeting of the American Society of Cell Biology and American Society of Molecular Biology and Biochemistry in 1989. Participants in the workshop compiled protocols that they use in their own labs and distributed their protocols as a handbook at that workshop. The handbook was written for researchers (graduate students, technicians, postdoctoral fellows, or lab supervisors) who have some familiarity with protein purification techniques, but who may not be aware of the particular details required in protein sequence analysis. This guide is an expanded and more complete version of that handbook. It is organized along the choices researchers face in deciding which approach to use in purifying their protein samples.

The introduction presents an overview of the protein sequencing reactions and the special reagents and protocols required in protein sequence analysis. Chapter 1 deals with the general strategies to consider in preparing an intact protein for N-terminal sequence analysis and for obtaining internal sequence from proteins with peptides generated by proteolytic and chemical cleavage. Chapter 2 describes HPLC methods to purify picomole (microgram) quantities of proteins and peptides. Chapter 3 describes sequence and amino acid analysis after purification by SDS–PAGE and blotting. For cases where the N terminus is blocked, Chapter 4 describes a practical method for digesting a blotted protein and obtaining N-terminal sequences of peptides eluted from the blot. Finally, for rapid confirmation of sequence, sequences of blocked N termini, or identification of modified amino acids, we describe mass spectrometry (Chapter 5). Sample data are also provided in each chapter for comparison. There is a reference section in the back of the guide in which references from the different chapters are grouped according to topic. Some of the references are not cited in the text, but are listed because they are generally useful for researchers who want to become acquainted with a particular procedure or method. Although this is somewhat unconventional, I hope it is helpful.

I wish to thank the authors of the chapters for contributing their collective experiences in protocol form, Ms. Pam Baud for helping to prepare this guide, Porton Instruments and the Millipore Corporation for sponsoring the ASCB/ASBMB workshop and handbook, and Dr. Phyllis B. Moses of Academic Press for helping to transform the handbook into a publishable guide.

Paul T. Matsudaira

Abbreviations

AARE	acylamino acid-releasing enzyme
ATZ	anilinothiozolinone
BNPS-skatole	2-(2-nitrophenylsulfenyl)-3-methyl-3-bromoindolenine
CAPS	3-[cyclohexylamino]-1-propanesulfonic acid
CI	chemical ionization
CID	collision-induced decomposition
CNBr	cyanogen bromide
DTT	dithiothreitol
EDTA	ethylenediaminetetraacetic acid
EI	electron impact
ES	electrospray
ES MS	electrospray mass spectrometry
FAB	fast atom bombardment
FAB MS	fast atom bombardment mass spectrometry
FPLC	fast-performance liquid chromatography
HPLC	high-performance liquid chromatography
IEF	isoelectric focusing

LSI MS	liquid secondary ion mass spectrometry
MALD-TOF MS	matrix-assisted laser desorption time-of-flight mass spectrometry
MS	mass spectrometry
MS/MS	tandem mass spectrometry
OPA	*o*-phthalaldehyde
PCR	polymerase chain reaction
PITC	phenyl isothiocyanate
PTC	phenylthiocarbamyl
PTH	phenylthiohydantoin
PVDF	polyvinylidene fluoride
rhM-CSF	recombinant human macrophage colony-stimulating factor
RP-HPLC	reversed-phase high-performance liquid chromatography
SDS–PAGE	sodium dodecyl sulfate
TCA	trichloroacetic acid
UV	ultraviolet

Introduction

Paul Matsudaira

*Whitehead Institute for Biomedical Research
and Department of Biology
Massachusetts Institute of Technology
Cambridge, Massachusetts*

Overview
Phenyl Isothiocyanate Degradation:
   The Edman Reaction
Identification of PTH-Amino Acids
Instrument Limitations
Chemical Limitations
Sample Loading Conditions

A Practical Guide to Protein and Peptide Purification for Microsequencing, Second Edition
Copyright © 1993 by Academic Press, Inc. All rights of reproduction in any form reserved.

NOTES

OVERVIEW

Automated protein sequencing instruments and methodology have evolved considerably in recent years with great improvements in sensitivity, speed, and ease of operation. Refinements in the 1980s include improved signal to noise and reduction in sequence by-product peaks by new coupling bases, delivery of lower concentrations of reagents, new sample cartridge designs, and detection of PTH-amino acids with photodiode array detectors. Current instruments can sequence 1–5 pmol of protein or peptide (1–18).

While the microsequencing instrument sets the lower limit in sensitivity, many microsequencing projects are limited by the ability to obtain protein/peptide samples in a form that is suitable for sequence analysis. The Edman chemistry imposes a number of constraints on the composition of protein/peptide samples. One has to be aware of these restrictions on sample composition and design purification procedures accordingly.

Most protein purification protocols utilize a combination of selective precipitation, ion-exchange chromatography, gel filtration chromatography, affinity chromatography, dialysis, and concentration steps in purifying a native and functional protein. The yields are measured by very sensitive antibody–enzyme assays, functional assays, and silver stain SDS–PAGE. As a result, nanogram quantities of protein are routinely isolated in a highly purified state. However, these samples are not always obtained in a form that can be directly submitted for sequence analysis. With milligram quantities of protein, it is relatively easy to change the composition of the protein solvent by dialysis or gel filtration and to concentrate the protein by lyophilization or ultrafiltration. However, microgram quantities of material are considerably more difficult to handle because peptides and proteins are especially susceptible to adsorptive losses at this level.

3

PHENYL ISOTHIOCYANATE DEGRADATION: THE EDMAN REACTION

Proteins are sequenced by degradation from their N terminus using the Edman reagent, phenyl isothiocyanate (PITC). The reaction (Fig. 1) is divided into three steps: coupling, cleavage, and conversion. During one cycle of this reaction, the N-terminal residue is removed from a polypeptide and identified by HPLC. The shortened polypeptide is left with a free N terminus that can undergo another cycle of the reaction. Today, most laboratories perform the Edman degradation with automated sequenators. The sample is commonly absorbed to a chemically modified glass fiber disc or electroeluted onto a porous polyvinylidene fluoride membrane (Fig. 2). The peptide remains bound to the support during the coupling and cleavage steps, which occur in a temperature-controlled reaction chamber. Instruments differ in whether the sample is covalently attached to the support and whether the coupling and cleavage steps are carried out in the gas or liquid phase.

In the coupling step PITC chemically modifies the α-amino group of the N-terminal residue to form a PTC-polypeptide. At pH 8.0–8.5 coupling is favored at α-amino groups. Coupling is inhibited by N-terminal modifications like acetylation, formylation, fatty acid acylation, and cyclization *(146–151)*.

In the cleavage step the PTC–N-terminal residue is rapidly cleaved from the polypeptide chain under anhydrous, acidic

Figure 1   The chemical reactions of the Edman degradation cycle. In the coupling step PITC modifies the N-terminal residue of a peptide. Acid cleavage removes the N-terminal residue as an unstable ATZ-derivative and leaves the shortened $(n-1)$ peptide with a reactive N terminus. The ATZ-derivative is converted in the last step to a stable PTH-amino acid.

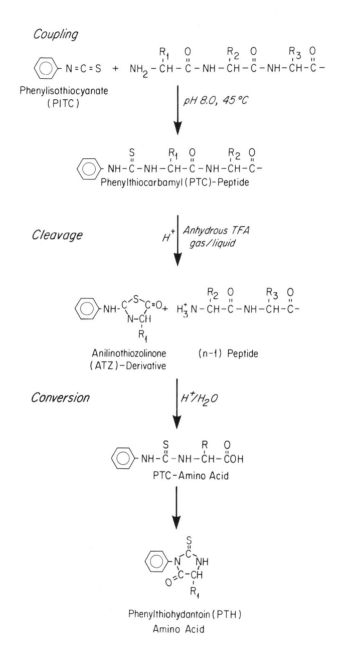

Coupling

Phenylisothiocyanate
(PITC)

pH 8.0, 45 °C

Phenylthiocarbamyl (PTC)-Peptide

Cleavage

Anhydrous TFA
gas/liquid

Anilinothiozolinone
(ATZ)-Derivative

(n-1) Peptide

Conversion

H^+/H_2O

PTC-Amino Acid

Phenylthiohydantoin (PTH)
Amino Acid

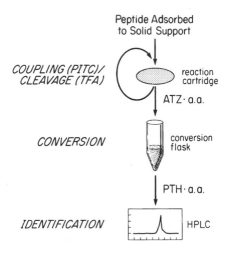

Figure 2   Different steps of a sequencing cycle are performed in separate parts of an automated sequencing instrument. The coupling and cleavage steps are performed on protein that is adsorbed or covalently coupled to a solid support located in the reaction cartridge. The ATZ-amino acid is washed into the conversion flask where it is converted into a stable PTH-derivative. The PTH-amino acid is injected onto an HPLC and the elution time of a peak identifies the amino acid.

conditions to liberate two products, an ATZ-amino acid and the $n-1$ polypeptide. The $n-1$ polypeptide has a reactive N terminus and can undergo another cycle of coupling and cleavage steps.

The ATZ-amino acid is extracted from the support and delivered to a small flask. ATZ-derivatives are unstable but can be converted in the third step of the cycle into stable PTH-derivatives. Conversion is a two-step reaction and occurs in aqueous, acidic solutions. First, the ATZ amino acid is rapidly hydrolyzed to a PTC-amino acid. In the second step, the PTC-amino acid cyclizes to a stable PTH-amino acid.

IDENTIFICATION OF PTH-AMINO ACIDS

The most convenient method for identifying the PTH-amino acids generated during each sequencing cycle is by UV absorbance and HPLC chromatography. All 20 amino acids are easily resolved by gradient elution from a reverse phase HPLC support. Each amino acid is detected by its UV absorbance at 269 nm and is identified by its characteristic retention time. A chromatogram of the PTH-amino acids is shown in Fig. 3. Depending on the instrument, some side products of the Edman degradation are also detected in the chromatograms. These are generated by reactions among PITC, DTT, and H_2) or base (trimethylamine, triethylamine, or diisopropylethylamine). Modified amino acids can be detected in some cases. For instance, phosphorylated serine is detected by the absence of a PTH-serine peak and the presence of a PTH-dehydroalanine peak. (During the sequencing cycle, the phosphate group is

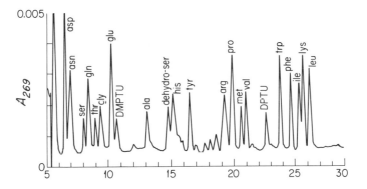

Figure 3   The elution pattern of the PTH-amino acid (15 pmol) standards. DMPTU and DPTU are by-products of the sequencing reaction and serve as useful reference peaks in the chromatograms.

7

hydrolyzed and serine is converted to a dehydroalanine deriva-
tive. The phosphate, if radiolabeled, can be extracted from the
support with 80% methanol in water.) Glycosylated amino acids
are also not identified directly. In most cases a glycosylated residue in the sequence generates a blank cycle because the PTH-
amino acid is not extracted from the sequenator for HPLC analysis. Most sequencing instruments are equipped for on-line
PTH analysis. In this mode, the PTH-derivatives are automati-
cally injected onto an HPLC system. Under the most sensitive
settings an HPLC can detect less than 1 pmol of each PTH-derivative. The sequence is determined by the appearance of a
PTH-amino acid in cycle x and the simultaneous disappearance
of that amino acid and the appearance of the next PTH-amino
acid (if it is different) in cycle $x + 1$. Chromatograms from the
first 15 cycles of sequence from a tryptic peptide of the F1
ATPase β subunit are shown in Fig. 4 of Chapter 4.

INSTRUMENT LIMITATIONS

The performance of the instrument is measured by:

1. The *sensitivity* limit of the instrument — the lowest amount
   of readable sequence (1 – 10 pmol).

2. The *initial yield* — a measure of the percentage of the sample loaded on the sequencer that is sequenceable. This is
   usually 50-80% unless the N terminus is blocked.

3. The *repetitive yield* — the percentage of sequence detected
   after each cycle of the Edman reaction. This value is measured by a linear regression fit to the amount of PTH-amino
   acid at each cycle. Extrapolation to cycle 0 gives the initial

amount of sequenceable peptide. The repetitive yield varies between 80 and 96% and depends on the sequence and size of the peptide.

These three factors partly determine how much sequence can be obtained from a sample. Longer sequences can be obtained from larger amounts of sample and higher repetitive yields (Fig. 4). The shaded areas in the figure highlight the amounts of PTH-amino acids that are recovered from 10- and 100-pmol amounts of starting material within a 90–95% range of repetitive yields. Theoretically, a 10-pmol peptide can be sequenced with a 95% repetitive yield for approximately 40 cycles before a limit of 1 pmol of sequenceable peptide remains. The same limit is reached for 100 pmol of peptide in 40 cycles with a 90%

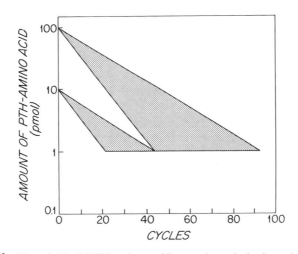

Figure 4 The yield of PTH-amino acid at each cycle is dependent on the repetitive yield. The amount of sequence that can be obtained is determined by the initial and repetitive yields. The shaded areas correspond to the amount of PTH-amino acids recovered when the repetitive yields are between 95% (top boundary) and 90% (bottom boundary). The sensitivity limit of the machine and the noise (background peaks) in the chromatograms determine the number of residues that can be identified.

9

repetitive yield. However, in practice, shorter lengths of sequence are obtained because background peaks generated by spurious cleavage of the peptide chain and by lagging sequences generated by incomplete cleavage or coupling reactions obscure the peak derived from the N-terminal residue. One could reasonably expect 10–15 cycles of sequence in most cases from 1–5 µg of sample.

CHEMICAL LIMITATIONS

The instrument limitations point out the reasons why sample preparation is so critical to successful sequence analysis. Contaminants that inhibit the Edman degradation should be eliminated prior to loading onto the sequencer. These include:

1. Destruction of or side reaction with PITC. Large amounts of primary amines will compete with the peptide for the Edman reagent, PITC, and will inhibit or prevent the coupling reaction. PITC is also destroyed by oxidizing agents.

2. Disturbance of proper coupling pH. Large quantities of buffer salts (with pK_a values <8.0) can alter the pH of the coupling step and induce inefficient coupling. Generally, salts are not soluble in the organic solvents used in the sequenator and remain on the sample support throughout the entire sequencing run.

3. Neutralization of cleavage acid (TFA). Buffers or other salts that remain on the sample support in large quantities may inhibit the cleavage reaction by neutralizing the trifluoracetic acid used in the cleavage step.

4. Reaction with the amino terminus of a peptide/protein. Contaminants in a sample can reduce sequencing yields by re-

acting with the amino terminus to produce a protein that is refactory to the Edman chemistry. Cyanate and aldehydes are common contaminants that may react with N termini of proteins. It should be noted that ammonium cyanate is generated in significant quantities in the concentrated alkaline urea solutions that are often used in protein purification.

5. Interference with the reverse-phase HPLC analysis of PTH amino acids. Any UV-absorbing (A_{269} nm) substance that is recovered from the sequencer along with the PTH-amino acids can complicate the reverse phase HPLC analysis. In addition, trace quantities of primary amines that react with PITC can produce UV-absorbing products that obscure PTH-amino acid peaks on chromatograms.

To avoid problems with the Edman chemistry the samples should be free of the following reagents:

1. *Buffers and primary amines.* Tris buffer is commonly used for protein purification. Tris and glycine are common in samples recovered from SDS–PAGE.

2. *Glycerol or sucrose.* These reagents are often added to buffers designed for the storage and handling of proteins. These compounds are not volatile and leave a highly viscous residue.

3. *Nonionic detergents.* Triton X-100, Brij, and Tween solutions often contain aldehydes, oxidants, and other contaminants that can inhibit the Edman degradation.

4. *SDS.* Large quantities of SDS can cause instrument malfunctions and may lead to the loss of sample from the filter.

NOTE Dialysis tubing is often a source of contaminants and other interfering substances. Avoid dialysis as a last step in sample preparation or use thoroughly cleaned, high-quality tubing. Do not dialyze the protein against water because the contaminants will remain adsorbed to the protein and dialysis membrane in the absence of a counterion salt or acid.

SAMPLE LOADING CONDITIONS

Most sequence analysis is performed using instruments in which the sequencing reactions and amino acid analysis are automated. The design and operating characteristics of these instruments require:

1. Sample volumes should be less than 150 µl. The sample is loaded onto a support, usually a glass fiber disk, that can only absorb 30 µl of liquid. Another aliquot of sample can be applied to the filter only after this volume has dried. Repeated loading and drying steps must be performed manually and can be quite annoying to the operator.

2. The sample should be in a volatile solvent or buffer such as acetic acid, formic acid, trifluoroacetic acid, triethylamine, acetonitrile, propanol, trimethylamine, water, or ammonium bicarbonate (if lyophilized repeatedly).

3. A minimum of 50–100 pmol of sample should be analyzed. The gas–liquid phase instruments can sequence 1 pmol of sample at their highest level of sensitivity. However, it is more practical to sequence larger amounts of protein to be confident of the sequence obtained or to be confident that the N terminus is blocked if no sequence is obtained. In most cases the amount of sequenceable material is underestimated by sample loss, inaccurate quantitation, or N-terminal blockage during sample preparation. Therefore, be sure to err on the side of too much rather than too little sample.

4. The sample can contain a small amount of detergent (<0.1% SDS), but larger amounts can wash the sample from the filter and cause bubbling in the tubing of the instrument.

5. All reagents and solvents must be of the highest purity (HPLC grade solvents, sequencing-grade and electrophoresis-grade reagents) available to avoid nonvolatile contaminants that may leave interfering residues upon drying. Avoid "molecular biology"-grade reagents; these often contain insoluble material and UV-absorbing contaminants.

One reason why the initial yields are often unexpectedly low is that the amount of sample present is overestimated. The most reliable quantitation method is from amino acid analysis. Other methods include micro-Lowry, BCA, or dye-binding assays, estimation from silver stain or Coomassie blue-stained gels, enzyme or antibody assays, or absorbance. The later methods are less accurate, especially in the low microgram amounts.

1 Strategies for Obtaining Partial Amino Acid Sequence Data from Small Quantities (<500 pmol) of Pure or Partially Purified Protein

Harry Charbonneau

Department of Biochemistry
Purdue University
West Lafayette, Indiana

Requirements for Protein Sequence Analysis
Purification Strategies
N-Terminal Analysis
Complete Fragmentation
Complete Fragmentation of Proteins Purified by
    SDS–Gel Electrophoresis
Limited Proteolysis
Computer Analysis of Protein Sequences
    Sequence Analysis Software
    Analyzing Database Searches Conducted with Partial
        Sequence Data
    Detecting Sequence Patterns or Motifs
    Designing Oligonucleotide Probes, PCR Primers, and
        Synthetic Peptide Antigens

A Practical Guide to Protein and Peptide Purification for Microsequencing, Second Edition
Copyright © 1993 by Academic Press, Inc. All rights of reproduction in any form reserved.

NOTES

16

REQUIREMENTS FOR PROTEIN SEQUENCE ANALYSIS

As described in the introduction, the key to obtaining protein sequence information is preparing a sufficient amount of pure material in a form that is compatible with the Edman chemistry. Currently, commercial instruments can routinely sequence at the 10-pmol level (0.5 µg of a 50-kDa protein) and under optimal conditions some can identify residues at the 1-pmol level. Therefore, many microsequencing projects are not limited by the sequenator, but by losses encountered as samples undergo final purification, concentration, and solvent exchange in preparation for sequence analysis. If internal sequence analysis is performed there will be additional losses associated with fragmentation and peptide purification. Uncertainties in sample recoveries during sample preparation and processing make it difficult to predict the amount of protein required for any given sequencing project. However, it is clear that the methods chosen for sample preparation have a major impact on both the amount of sample required and the overall success of a microsequencing project. This chapter provides an overview of the available options by outlining in flow charts the general strategies available for N-terminal and internal sequence analysis. The strategy to follow should be determined by (1) the amount of sample, (2) if impure, the number of components in the crude sample, (3) the molecular weight of the protein, (4) one's relative familiarity with electrophoresis/electroblotting and HPLC methods, and (5) whether internal sequences are needed. An important point to consider is that all the purification methods described in this guide employ denaturing conditions. This simplifies the complexity of the purification by reducing the number of steps involved and by ensuring that sequence will be obtained from a single polypeptide.

PURIFICATION STRATEGIES

Chromatography and SDS gel electrophoresis (*9, 50–60*) are two general approaches for purifying a protein from a complex mixture. With 50–1000 μg of material, the reverse phase, ion exchange, and gel filtration modes of HPLC and FPLC have proven to be the methods of choice for achieving final purification because of their speed and resolving power. FPLC columns generally have higher capacity than most HPLC columns and are better suited for use with larger quantities of protein. In contrast, most HPLC systems operate on a smaller scale with greater sensitivity and speed. In addition, with HPLC methods, it is often possible to use volatile buffers that will not interfere with subsequent sequence analysis. Because recoveries of some proteins from reverse-phase HPLC columns can be low, it is best to perform preliminary analytical runs using a small aliquot sample to estimate recoveries prior to separating larger amounts of sample. Because FPLC systems operate at lower pressure, the columns and associated tubing are made of plastic or glass which are thought to give better recoveries than the stainless steel materials required for HPLC. Small quantities (nanograms) of protein can be purified using new generation FPLC systems (SMART system, Pharmacia), but the salt and buffer used in the chromatography must be removed before sequence analysis. Prior to sequence analysis, column fractions should be analyzed by SDS–PAGE to identify peaks containing the protein of interest and to verify its purity.

If possible, the investigator should utilize a final purification step that yields a protein sample in a solvent compatible with sequence analysis (see the introduction). The additional manipulations required to place the pure sample in an appropriate solvent will often result in significant losses particularly when small amounts of protein (<500 μg) are manipulated. With microgram quantities, it is often best to precipitate the protein with trichloroacetic acid or organic solvents and to redissolve it in an appropriate solvent. If recoveries are sufficiently high, reverse-phase HPLC is an excellent desalting method. With milligram quantities, buffer exchange is trivial and can be accomplished by gel filtration or dialysis.

With the development of effective techniques for sequencing and digesting proteins separated by gels, one- or two-dimensional electrophoresis has become an excellent alternative approach for protein purification in microsequencing projects. Electrophoresis using one- or two-dimensional gels is unsurpassed in separating a mixture of proteins into individual components. In many cases, one-dimensional SDS–gel electrophoresis of a crude protein mixture can achieve a level of purity that might require multiple steps using standard chromatographic procedures. The use of SDS–PAGE for sample purification has four major advantages: (1) it is simple and fast, (2) it permits the analysis of partially purified samples, (3) it eliminates the losses associated with complicated desalting or buffer exchange steps, and (4) it is a commonly used technique that can be performed in most laboratories. Hence electrophoresis-based strategies (Figs. 1 and 4) are ideal for the final purification and preparation of microgram quantities of protein.

N-TERMINAL ANALYSIS

N-terminal sequence is extremely useful for molecular cloning studies. The goal of most gene cloning efforts is to obtain the complete coding region of the protein. With a probe to the N terminus of the coding region, a researcher will be able to isolate full-length clones from libraries constructed by priming from the poly (A) tail of the mRNA. The N-terminal sequence also identifies the beginning of the protein, which for processed proteins or peptides may not be obvious by inspecting a sequence obtained by gene cloning. For N-terminal sequence analysis of most samples, we prefer to use SDS–PAGE for sample preparation. The advantages of this approach were noted previously. As outlined in Fig. 1, we prefer to electroblot samples onto PVDF membranes *(65, 83–90)*, stain with Coomassie blue, excise the band, and perform standard gas phase sequence analysis. The critical step in this approach is to adjust the electroblotting conditions for complete transfer. The details for this electroblotting protocol are given in Chapter 3. Alternatively, proteins can be stained in the gel, the band excised from the gel, and the protein electroeluted into solution *(76–79)*. After elution the protein must be concentrated, separated from gel-derived contaminants, and exchanged into a volatile buffer that is appropriate for sequencing. Electroblotting is preferred over electroelution because fewer manipulations are required. With current electroblotting protocols, N-terminal sequence is usually obtained from the electrophoresis of approximately 10–100 pmol of protein.

N-terminal sequence analysis of high-molecular-weight proteins (>100 kDa) presents special problems. Proteins of this size are more difficult to electrophorese out of a gel and are usually not amenable to strategies based on SDS–PAGE. High-molecular-weight samples are usually loaded directly into the sequenator, but they must be purified and exchanged into appropriate solvents using standard procedures which generally require larger quantities of starting sample. It is important to note that the number of residues that can be identified from N-terminal sequence analysis of large proteins is generally limited due to rapid accumulation of interfering background sequences. With limited amounts of a large protein, all of the sample should be used for

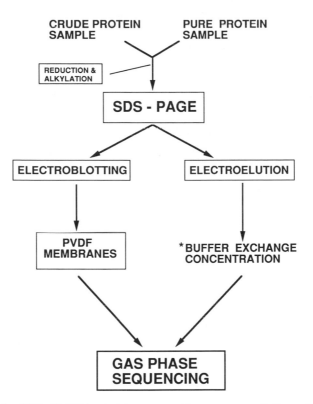

Figure 1 SDS–PAGE is used to prepare either pure or partially purified proteins for N-terminal sequence analysis. The protein can be electroblotted onto a membrane or electroeluted into solution and exchanged into a solvent suitable for sequencing.

internal sequence analysis (see below) rather than N-terminal sequencing which often yields little or no sequence.

Failure to obtain N-terminal sequence is usually attributed to N-terminal blockage or insufficient quantity of sample. Chapter 3 describes methods for chemically or enzymatically removing blocked N-terminal residues from electroblotted samples.

COMPLETE FRAGMENTATION

It has been estimated that greater than 75% of the intracellular proteins from eucaryotic cells have blocked amino termini as a result of post-translational modification and are refractory to the Edman degradation *(146–148)*. In addition, purification or gel electrophoresis of microquantities of proteins can result in artifactual blocking of their N termini. Thus sequence data can only be obtained either by removing the blocking group (Chapter 3) or by fragmenting these proteins by complete digestion or limited proteolysis (Figs. 2–5).

Complete fragmentation generates multiple peptides and offers the potential to obtain more extensive sequence information. This is an advantage in molecular cloning projects as it provides more alternatives for designing oligonucleotide probes or PCR primers and facilitates attempts to verify the authenticity of putative positive clones. Even if N-terminal sequencing is successful, generation of additional internal sequence is desirable in most projects.

For complete fragmentation, the protein is digested at specific residues using site-specific proteases or chemicals as listed in Table I. The choice of cleavage reagent depends primarily on the number and/or size of fragments desired. Cleavage at high frequency residues (e.g., Lys or Arg) increases the potential for determining more internal sequence but makes subsequent purification steps much more difficult due to the relatively large number of peptides obtained. Conversely, targeting less abundant residues (e.g., Met or Trp) simplifies purification but reduces the amount of internal sequence that can be obtained. In our laboratory, we prefer cleavage at Lys with *Achromobacter* protease I *(41)* because of the increased potential for internal

Table I

Cleavage Reagents Commonly Employed in Microsequence Analyses

Cleavage agent	Specificity[a]	Source	Supplier[b]	References
Endoproteinase Lys-C	Lys-x	Lysobacter enzymogenes	Bm	(44)
Achromobacter protease I	Lys-x	Achromobacter lyticus	Wa	(41)
Endoproteinase Arg-C	Arg-x	Mouse submaxillaris gland	Bm, Ta	(35)
Endoproteinase Asp-N	x-Asp, x-cysteic acid	Pseudomonas fragi	Bm, Cb	(40)
Asparaginylendopeptidase	Asn-x	Jack bean	Ta	(46)
V8 protease	Glu-x, Asp-x	Staphylococcus aureus	Bm, Pi	(34)
Trypsin	Lys-x, Arg-x	Bovine pancreas	Bm, Pi	(36, 32)
Chymotrypsin	Trp-x, Tyr-x Phe-x, Leu-x Met-x	Bovine pancreas	Bm	(32)
Cyanogen bromide	Met-x		Ek, Pi	(30)
BNPS-skatole	Trp-x		Pi	(33)
Iodosobenzoic acid	Trp-x		Pi	(39, 43)
N-Chlorosuccinimide	Trp-x		Al, Ek	(37)

[a] The peptide bond cleaved lies between the designated amino acid and the "x".

[b] The vendors listed are not necessarily the exclusive source for these reagents. Suppliers are indicated by the following abbreviations: Bm- Boehringer Mannheim; Pi, Pierce Chemical Co.; Cb, Calbiochem; Ek, Eastman Kodak Laboratory Chemicals; Al, Aldrich Chemical Company, Inc.; Wa, Wako Chemicals USA, Inc.; Ta, Takara Biochemical, Inc.

sequence; if larger fragments and simpler peptide separations are desired CNBr cleavage at Met *(30)* is the method of choice. The purification of peptides from a complete digest is not trivial and generally requires HPLC chromatography. Chapter 2 describes procedures for the enzymatic digestion of proteins and the reverse-phase HPLC purification of peptides.

For large amounts of protein (≥2 nmol) one should follow the guide depicted in Fig. 2. In most cases, cysteine residues should be reduced and alkylated prior to digestion to cleave any disulfide bonds and prevent them from reforming. This ensures that the protein sample remains completely unfolded for efficient cleavage and eliminates formation of disulfide-linked

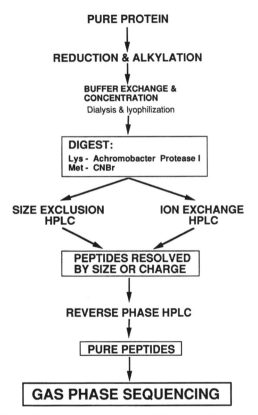

Figure 2 Internal sequence is obtained by complete fragmentation of 0.1–2.0 nmol of pure, reduced, and alkylated protein. Peptides generated by digestion can be prefractionated by size or net surface charge prior to the final revese-phase HPLC purification. The preliminary fractionation greatly simplifies the subsequent reverse-phase step.

peptides, which complicates subsequent HPLC separation and sequence analysis. In addition, unmodified cysteine residues cannot be positively identified during the Edman degradation unless they have been converted to more stable derivatives. Thus, the reduction and alkylation of protein samples is usually advantageous. Protocols for reduction and carboxymethylation or carboxyamidomethylation *(26,78)* with iodoacetic acid or iodoacetamide, respectively, and pyridylethylation *(104)* with 4-vinylpyridine have been described. The need to remove excess reagent and unwanted by-products can lead to sample loss and is one drawback of cysteine modification. However, with greater than 2.0 nmol of sample, this can often be accomplished with acceptable yields using size exclusion or reverse-phase HPLC chromatography. As outlined in Fig. 2, the purification of peptides from digests of large proteins ($\geq$40 kDa) is usually facilitated by using a two-step procedure. Use of a size exclusion or ion exchange HPLC step prior to reverse-phase chromatography reduces the number of components that must be resolved and increases the chances of obtaining single sequences upon analysis of peaks. Because peaks obtained from a single HPLC separation often contain more than one peptide, it is best to check the purity of the fraction prior to sequence analysis using microbore HPLC (1.0–2.1 mm columns), capillary zone electrophoresis, or mass spectrometry (laser or plasma desorption, electrospray, etc.).

With smaller quantities of sample (0.1–2.0 nmol), the fragmentation and peptide purification procedures must be streamlined in order to minimize losses. As illustrated in Fig. 3, digests are resolved by a single reverse-phase HPLC step without the intermediate size or charge separations. To avoid losses associated with desalting samples at the sub nanomole level, reduction and alkylation reactions can be performed after digestion *(104)*

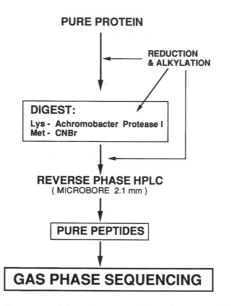

Figure 3 The digest resulting from complete fragmentation of picomole quantities of protein is separated directly by reverse-phase HPLC. The prefractionation steps are eliminated to avoid losses. Reduction and akylation at the time of the digest or immediately after digestion minimize sample manipulation and also prevent losses.

so that excess reagents and by-products are removed during reverse-phase HPLC purification of peptides. Alternatively, procedures have been developed (see Chapter 2) for performing the reduction and alkylation reaction and enzymatic digest in the same vessel without the need for an intervening desalting step. It should be noted that many enzymatic digests can be performed in the presence of limited amounts of detergents, denaturants, or organic solvents *(48)*. This facilitates the fragmentation of small quantities of proteins that display poor solubility in aqueous solvents or are resistant to enzymatic cleavage. Furthermore, with membrane proteins, it may not be necessary to undergo the complications and losses associated with the removal of detergents prior to digestion.

COMPLETE FRAGMENTATION OF PROTEINS PURIFIED BY SDS GEL ELECTROPHORESIS

As noted above, SDS electrophoresis is a simple yet powerful method for purifying small amounts (<1 nmol) of protein *(61)*. Thus, if the protein of interest can be linked to a specific band on an SDS gel (e.g., by Western blotting), it is possible to generate an internal sequence despite starting with an impure sample. As outlined in Fig. 4 and in Chapters 3 and 4, there are

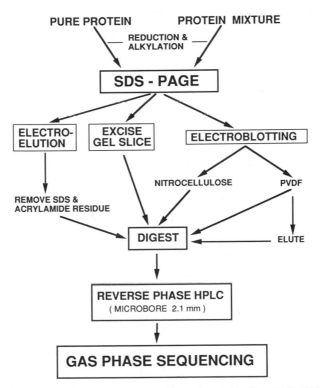

Figure 4 Picomole quantities of protein can be purified by SDS–PAGE and recovered by electroelution or electroblotting. The recovered sample is then digested and purified by narrowbore (2.1 mm ID) reverse-phase HPLC. Alternatively, the sample can be digested in an excised gel slice.

several approaches for obtaining an internal sequence from proteins purified by SDS–PAGE. One can either: (1) digest the sample within a gel slice, (2) digest the sample after electroelution from SDS gels, or (3) digest the sample after blotting onto nitrocellulose or PVDF membranes.

Digestion of a protein within a stained gel slice generates unwanted by-products that may complicate subsequent purification by HPLC and result in added manipulations. Most of these problems are overcome by using SDS–PAGE for separation of cleavage products. In this method, the excised gel band is digested and then subjected to a second SDS–PAGE step; in some cases, it is possible to load the digested gel slice directly into the well of the second gel to minimize sample transfer and handling. Fragments separated on the second gel are electroblotted onto PVDF membrane and sequenced. Generally, complete fragmentation generates small fragments that may not be resolvable by a second SDS gel or may not bind efficiently to the membrane during the blotting step. Thus, it is advisable to use chemical cleavages that target low frequency residues such as Trp or Met. Enzymatic digests (e.g., V8 protease) that usually do not produce a limit digest under these conditions are also recommended. Use of a tricine–SDS system *(70)* instead of the common Laemmli system *(62)* makes this approach more attractive because fragments within a lower molecular mass range (5–10 kDa) can sometimes be resolved and blotted.

In alternative approaches, the protein is digested after electroelution from the gel or after electroblotting onto a membrane. Chapter 3 describes the digestion and purification of samples electroeluted from gels. *In situ* cleavage on PVDF or nitrocellulose membranes is described in Chapters 2 and 4. Peptides generated from *in situ* digests are generally purified by narrowbore HPLC methods. However, if large fragments are expected (≥10 kDa), it may be advantageous to re-electrophorese the digests and blot to a second PVDF membrane for sequencing (see

Chapter 3). With all of these procedures, samples should be reduced and alkylated prior to SDS–PAGE. This can be accomplished with little or no loss of sample since the reaction can be performed in the SDS sample buffer just prior to electrophoresis as described *(78, 104)*. Because of the added steps involved with digestion and peptide purification, these procedures typically require 2–5 times more starting sample than is needed for direct N-terminal sequencing from gels.

LIMITED PROTEOLYSIS

Many large proteins are composed of multiple domains that are compact units of structure believed to have the ability to fold independently. Typically, each domain is responsible for a specific function (e.g., catalysis, allosteric regulation, metal binding, etc.). Segments connecting domains are often exposed loops that are protease sensitive, whereas the domain core is protease resistant because of its compact structure. Thus, with a small ratio of protease to substrate (e.g., 1:100 or 1:1000), multidomain proteins are only cleaved at the most accessible sites, usually loops between domains. Limited proteolysis *(45)* generates peptide fragments of 10–40 kDa in size that can be used to obtain internal sequence (Fig. 5). Table I lists the proteases that are commercially available in sequencing grade purity; most have been used for limited proteolysis. The most effective protease and the optimal conditions for its use must be determined empirically in separate experiments that are most easily analyzed by SDS–PAGE *(45)*. In these experiments, several parameters should be optimized: ratio of protease to substrate, reaction temperature, and incubation time. It is important that the reactions be stopped effectively and reproducibly using

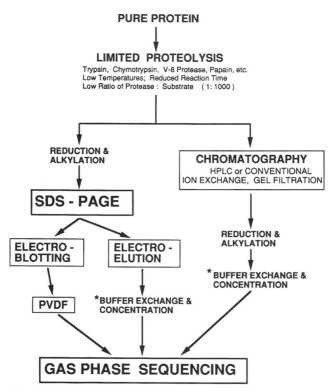

Figure 5 Limited proteolysis of pure protein generates large peptides that can be purified by either SDS–PAGE or chromatographic procedures. SDS–PAGE-purified samples are electroblotted or electroeluted prior to sequencing.

either protease inhibitors or acid precipitation. Inhibition of the protease is particularly important if the reaction is to be analyzed by SDS– PAGE (45). It is also important to use proteases of the highest purity because contamination by other proteases will generate additional peptides by cleavage at other sites.

After limited proteolysis, subnanomole digests are prepared for sequence analysis by blotting or electroeluting fragments from SDS gels (Fig. 5). As mentioned previously, electroblotting to PVDF membranes is the preferred method in our laboratory. With larger quantities of starting sample, HPLC or FPLC may also be used to purify fragments for sequencing. If there is sufficient sample, use of nondenaturing solvents in conjunction with FPLC or conventional chromatography will permit an assessment of the biological properties of isolated fragments. Although limited proteolysis generates a small number of fragments and limited internal sequence, it usually yields peptides that can be purified by SDS–PAGE and provides information about the domain organization of the protein that may be relevant when its complete sequence is ultimately revealed by molecular cloning. When sufficient material is available, it is sometimes advantageous to perform both limited proteolysis and complete fragmentation.

COMPUTER ANALYSIS OF PROTEIN SEQUENCES

The most common objectives in microsequencing determinations are to provide partial sequences for (1) design of PCR primers and oligonucleotide probes, (2) identification of a purified protein or an "interesting spot or band" on a gel, and (3) preparation of anti-peptide antibodies. Computers are essential for conducting the database searches that permit identification of new sequences or recognition of their homologs. While they are not always required, computers facilitate the selection of sequences to be used as a basis for designing probes or preparing peptide antigens.

The data obtained from microsequencing projects vary greatly but usually include several noncontiguous and nonoverlapping segments of sequence that together account for only a small percentage of the total residues. Although these fragmentary data impose certain limitations, it is important that the

31

investigator use the new sequence data to conduct a computer-assisted search of the sequence databases before proceeding with the project. Such searches can provide valuable insights that will influence how projects are carried out. Even a limited and highly fragmented set of partial sequences can be utilized to detect identical sequences and, in most cases, one can usually determine whether the protein under investigation has already been sequenced. In fact, as noted earlier, identification of an unknown protein sample is the major goal in many projects. While they are not always conclusive, database searches may also provide the first suggestion for the function of an unknown protein by revealing potential homologs. Homologous proteins have evolved from a common ancestor *(152)*, have similar three-dimensional structures, and in most cases, perform similar functions.

This section provides a brief description of procedures for analyzing protein sequence with an emphasis on the use of the fragmentary data anticipated from microsequencing procedures outlined in this manual. For a more detailed discussion, the reader is referred to several excellent articles and books *(152–157)* on this topic. The primer by Doolittle *(155)* is recommended as an excellent starting point for the novice.

Sequence Analysis Software

There are numerous software packages that can assist the investigator in the analysis of protein and nucleic acid sequences. Most packages include programs for (1) searching sequence databases for similar or homologous entries, (2) aligning sequence pairs, (3) statistically evaluating the significance of alignments, (4) detecting specific sequence motifs or patterns, (5) translating a nucleic acid sequence into a protein sequence, (6) predicting antigenicity, and (7) predicting secondary structure. Most of these programs can be utilized in the analysis of data obtained from microsequencing projects (see below). Sequence analysis packages are user-friendly and readily learned; however, it is advisable for the novice to seek assistance from an experienced user, particularly when analyzing the computer output. Packages are available for operation on both mainframe machines and laboratory PCs. Software from the University of Wisconsin Genetics Computer Group (GCG), the National Biomedical Research Foundation (NBRF), and Intelligenetics Inc. are widely used on VAX computers. Numerous packages are also commercially available for both the IBM and Macintosh personal computers. These packages and their programs are far too varied and change too rapidly to warrant a detailed discussion here; the reader should consult von Heijne *(154)* and Gribskov and Devereux *(158)* for a more complete listing of software packages. Although there are many, the most commonly used protein databases *(158)* are the Protein Identification Resource (PIR, from the National Biomedical Research Foundation) and the SWISS-PROT (Department de Biochimie Medicale, C.M.U., Switzerland). The GenBank and EMBL are two widely distributed nucleic acid databanks.

Virtually all of the sequence analysis packages include one or more programs that search protein databases for entries having similarity to a query sequence. Most search programs are also capable of searching nucleic acid databases with a protein sequence by first translating the database entries in all possible reading frames prior to comparison with the query sequence. A brief description of search programs and their use in the analysis of a new sequence follows. Although several different strategies and algorithms are commonly employed *(155–159)*, most search programs generate similar output—a list of database entries having segments that are either identical or similar to the test sequence. The degree of similarity between paired sequences is usually expressed by a score that is based on the number of identical residues or a scoring matrix that assigns high values to amino acid replacements that occur frequently during evolution. Generally, the program ranks database entries according to their similarity scores with the test sequence. However, the top-scoring entries do not necessarily display significant similarity to the test sequence. In order to provide evidence that similarities are significant and therefore indicative of a homologous relationship *(152)*, the investigator must first analyze the top-scoring pairs from the search program output using a separate alignment program. As the name implies, these programs are designed to extend and optimize alignments between sequence pairs. With distantly related sequences, this optimization usually requires insertion of gaps. After optimization, the alignment is evaluated statistically and is assigned a score that indicates whether the observed similarity occurred by chance or by evolution from a common ancestor *(156,157)*. Programs generating dot matrix plots may also be useful for examining and evaluating a potential structural relationship among sequence pairs. Common ancestry is likely if two long sequences ($\geq$100 residues) display more than 25% identity *(155)*. Establishing a homologous relationship for sequences having less similarity is much more difficult; in these borderline cases, the investigator must also consider the biological properties of the protein in question in addition to statistical parameters.

Analyzing Database Searches Conducted with Partial Sequence Data

Editing the Dataset

Before initiating a search, all sequences of poor quality or of questionable reliability should be eliminated from the dataset. Usually, this is done by the sequenator operator who screens the data before passing it on to the investigator for further analysis.

The Edman degradation of a sample composed of multiple peptides is not interpretable and such data do not warrant further
analysis unless one sequence clearly predominates and is easily
discernable from all others. Sequences containing only a few
scattered, unassigned positions can often be used for searches,
but those with extended gaps or a large number of tentative assignments should be excluded. Next, it is best to screen the
dataset for any contaminating sequences that may be derived
from either proteases used for cleavages or protein reagents
(e.g., immunoglobulins, bovine serum albumin) employed in
the preparation and purification of the sample. Contamination
by protease-derived fragments is most likely when conducting
in situ cleavage of electroblotted samples, but should be considered as a possibility in all microsequencing projects. In fact, use
of a protease with a known sequence is recommended so that any
contaminating fragments are readily recognized. The complete
sequences of trypsin, chymotrypsin, subtilisin, thermolysin,
Lysobacter endoproteinase Lys-C, and *Achromobacter* protease
I are known (see Appendix), whereas partial sequences for clos-
tripain and *Staphlycoccus aureus* V8 protease are available. It
is useful to gather sequences of the other common protein contaminants into a single computer file that can be rapidly scanned
before conducting a complete database search. Having excluded
all sequences known to be derived from contaminants or of poor
quality, one is ready to scan an entire protein database. Although
there are redundancies between protein and nucleic acid databases, at least one of each kind of database should be searched.

Establishing Homologies from Database Searches

Initially, search results should be examined for the presence of
a database entry that may have a sequence identical to that of
the sample. Generally, identity between the sample and a data-

base entry is readily recognized since all segments searched should exhibit either absolute identity or a high degree of similarity ($\approx 90-99\%$ identity) to corresponding regions from only one sequence in the database. When examining the results, keep in mind that even though an unknown sample and database entry may actually have the same sequence, some of the sample sequences may not exhibit absolute identity because of errors and/or gaps that occur with significant frequency during microsequence analysis. With limited data, deciding whether a high degree of similarity is sufficient to indicate identity may be difficult and will depend on judgements about the likelihood of sequencing errors at nonidentical positions. If the sample data consists only of short segments containing relatively few total residues, establishing identity to other sequences can present problems since a match between two short segments (≤ 8 residues) is not necessarily an indication that the two proteins are identical, particularly if the database sequence is relatively long *(161)*. However, as noted above, if all such short sequences match only one database entry then there is a much greater chance that the sequences will prove to be identical. If segments within the dataset display identity to more than one entry from the database, an impure starting sample is indicated. Heterogeneity should also be considered when a single database entry appears to be identical to some but not all sequences from the dataset. It is important to remember that no matter how extensive it may be, partial sequence data cannot prove that a sample and database entry are completely identical since some proteins differ in only a few residues that are localized to one short segment.

If the sample sequences appear to be unique, then top-scoring matches from the search program output should be examined for database entries that may be potential homologs. The longer segments from the dataset will be the most useful for detecting possible structural relationships. With long segments (≥ 25 residues), decisions about whether there is a homologous relationship to the top-scoring database entries can be based on statistical parameters generated with alignment programs. If there are segments of this length included in the data, it may be possible to build a strong case for a homolog. However, segments of 10–20 residues are more common. Searches conducted with sequences of this length generally do not provide sufficient evidence for establishing homologous relationships, but may permit the identification of database entries that are similar enough to be viewed as potential homologs. When comparing short sequences (<20 residues), the chance for random matches increases dramatically, making it more difficult to establish a homologous relationship *(155)*. For instance, identical sequences of 7 or 8 residues have been found in unrelated proteins with a surprisingly high frequency *(160)*. With segments in this size class, the strongest argument for a homologous relationship is obtained when one database entry (or proteins within a single family) consistently exhibits a high degree of similarity to most of the sample sequences.

Another problem exists because distantly related proteins (<35% sequence identity) typically have segments of low or undetectable similarity (up to 20–30 residues) intermixed with those of high similarity. If by chance the newly determined sequences are derived from regions of low similarity, it is possible that an authentic homolog will be missed in a search. Of course this problem is minimal if the partial sequence includes relatively long segments (>20 residues) and a large number of total residues. Data from microsequence analyses, particularly that obtained near the end of sequence runs, sometime contains errors and unidentified residues that further complicate interpretation. When evaluating search results, it is important to consider the potential for errors. This usually requires consultation with the sequenator operator who can point out tentative residue assignments or positions where errors are most likely.

Harry Charbonneau

Large Proteins Are Often Chimeric in Function and Sequence

When scanning the search program output, it is also important to remember that many large, multifunctional proteins are chimeric. Chimeric proteins appear to be products of a fusion between two distinct genes and display homology to two or more distinct proteins. If new sequences are derived from such a chimeric protein, then two or more different database sequences may appear to be related to sequences derived from the same sample. However, this type of pattern may also result from an impure starting sample.

Scanning the data for the presence of sequence motifs or patterns (see below) may also provide evidence regarding a possible homology since consensus sequences are thought to be a characteristic of specific protein families. Information about the chemical and functional properties of the sample being studied provides valuable insight that can often indicate whether a potential structural relationship merits further consideration. In summary, one must exercise caution and avoid overzealous interpretation when analyzing database searches conducted with partial sequence data.

Results from a collaborative microsequencing project *(161)* illustrate how early database scans were interpreted and how they influenced the direction of our project. Complete fragmentation of 800 pmol of a newly discovered calcium-dependent plant kinase *(161)* ultimately yielded three unambiguous and unique sequences of 33, 19, and 13 residues in length. Other sequences of questionable reliability were not analyzed further. A search of the PIR database indicated that the two longest sequences might be related to proteins from the kinase family since in both cases the majority of the top 20 scoring PIR entries in the output were protein kinases. However, the shortest sequence of 13 residues was not similar to protein kinases. The longest sequence was subjected to further analysis using a program to optimize its alignment to a corresponding 35 residue

38

segment from the cGMP-dependent protein kinase. This alignment clearly indicated a homologous relationship since the sequence identity was 29%, and a statistical evaluation *(157)* showed that the probability of such an alignment occurring by chance was $<10^{-16}$. No attempt was made to align and score the shorter 19 residue sequence. This analysis (1) showed that our sequences were unique in that they had not been reported to the databases, (2) confirmed that the protein was homologous to other protein kinases, and (3) suggested the best sequences for use in designing oligonucleotide primers for PCR. Using PCR primers based on these sequences, a full-length cDNA was eventually cloned and sequenced *(161)*. The translated, full-length sequence showed unequivocally that the plant protein was a member of the protein kinase family.

Detecting Sequence Patterns or Motifs

Many protein sequences contain patterns or motifs that are believed to be hallmarks of structures involved in particular functions. Often these segments are thought to be binding sites, key catalytic residues, or special recognition sites for intracellular processing or post-translational modification. A well known example is the Gly-Xaa-Gly-Xaa-Xaa-Gly pattern that is found in certain nucleotide-binding domains. Because the presence of these motifs can give valuable clues about the function of a newly discovered protein, all new sequences should be screened for their presence. If the new dataset includes a large number of residues, such screens are best done with the assistance of computer programs. Some software packages have specialized routines for detecting patterns; if not, the motifs of interest can usually be entered as a query sequence and used with searching programs to scan new sequence data. Because they are usually short, random matches to motifs can appear with high frequency and judging their significance is often difficult *(155, 156, 160)*. The functions and properties suggested by their presence must ultimately be proven by direct biochemical experimentation.

Designing Oligonucleotide Probes, PCR Primers, and Synthetic Peptide Antigens

An antibody specifically recognizing a newly discovered protein can be an extremely important reagent that can be used for its isolation, identification, and characterization. One approach for generating antibodies to a protein that has not yet been purified is to use a synthetic peptide corresponding to the appropriate internal sequences as an antigen. The synthetic peptides are coupled to appropriate carrier proteins and used for production of antibodies that are usually capable of specific recognition of the full-length protein. There are many factors to consider in designing such peptides, one of the most important is the likelihood that the sequence is on the protein surface and is an antigenic determinant. It is also necessary to consider whether the peptide will be soluble in an aqueous solution and therefore easily manipulated. Predictions of the antigenicity of sequences are based on scales (e.g., Kyte-Doolittle, Hopp-Woods) that rank amino acids according to their polarity or hydrophilicity *(162, 163)*. The average polarity of sequences are plotted with the assistance of programs that are found in most software packages; usually the user can choose from two or three different polarity scales. Most investigators use peptide antigens of 10 or more residues; shorter sequences within the dataset are usually not considered. Unless the partial sequence data contain a large number of total residues, a computer is not required to screen for antigenicity; manual inspection using a table of hydrophilicity values for each amino acid may suffice.

The design of ideal oligonucleotide probes and primers for the PCR reaction requires that the protein sequence be reverse translated so that regions of minimal codon redundancy can be recognized. With a large set of relatively long partial sequences, this is most easily accomplished with the assistance of computer programs that score sequences for probe redundancy. Although visual inspection is trivial when fewer residues are involved, use of a computer can reassure the user that ideal sites have not been inadvertently overlooked. Of course, it is imperative to exclude all sequences of poor quality when designing oligonucleotide probes and PCR primers.

2 Enzymatic Digestion of Proteins and HPLC Peptide Isolation

Kathryn L. Stone and Kenneth R. Williams

Howard Hughes Medical Institute, and W.M. Keck Foundation
Biotechnology Resource Laboratory
Boyer Center for Molecular Medicine
Yale University
New Haven, Connecticut

Introduction
Sample Preparation for Enzymatic Digestion
    Preparation of Non-SDS Containing Samples
    Preparation of SDS Gel-Eluted Proteins
    Proteins Isolated by Electroblotting onto PVDF
        Membranes
    Proteins Isolated by SDS–PAGE for in-Gel Digestion
Enzymatic Digestion of Proteins
    Trypsin, Chymotrypsin, and Endoproteinase Lys-C
    Digestion Procedure
    In Situ Enzymatic Digestion in SDS–PAGE Gels
HPLC Fractionation of Subnanomole Amounts of
    Enzymatic Digests
    Notes on Reverse-Phase HPLC
Summary of Results from the Cyanogen
    Bromide/Acetonitrile Elution and Tryptic
    Digestion of 26 "Unknown" Proteins

NOTES

INTRODUCTION

Many eukaryotic proteins have blocked amino termini *(146–148)*. Often, therefore, in order to obtain amino acid sequences suitable for generating oligonucleotide probes it is necessary to cleave the protein either enzymatically or chemically and to then isolate and sequence one or more of the resulting fragments. Since SDS–PAGE is often the final purification step, those procedures for obtaining internal amino acid sequences that are compatible with samples obtained via SDS–PAGE are the most useful. As described in this handbook, there are at least three different alternative strategies for generating and isolating peptides from SDS–PAGE-purified proteins. First, the protein can be electroeluted from the gel and then digested (described in Chapter 3); second, the protein can be digested in the gel matrix prior to eluting the resulting peptides; and third, the protein can be electroblotted onto a membrane. In the last instance the protein can either be enzymatically digested on the membrane (described in Chapter 4) or be eluted from the membrane and then subjected to enzymatic digestion. Each of these approaches has its own advantages and disadvantages and, so far, no single procedure appears to be clearly superior to the others. While blotting onto PVDF provides an easy means of removing salts and detergents from proteins destined for internal amino acid sequencing, the relatively low blotting efficiencies that can be routinely obtained (i.e., we obtained an overall average PVDF blotting efficiency of 34% for 11 proteins that ranged from 14 to 116 kDa) is a disadvantage. While the "in-gel" digest approach avoids losses resulting from blotting onto membranes, losses may be encountered with this approach while the SDS detergent is being allowed to diffuse out of the gel matrix (prior to enzymatic cleavage). Because of the reliability of PVDF blotting/washing in removing SDS and the ease with which PVDF-blotted proteins can be eluted in high yield in volatile solvents (following *in situ* cleavage with cyanogen bromide as described below), we often recommend this approach. Since the resulting cyanogen bromide fragments are isolated in volatile solvents, they can be easily reconstituted in the minimal volumes of 8 *M* urea that are essential to obtain the comparatively high substrate concentrations required for complete enzymatic digestion. However, for those proteins that have very low blotting efficiencies (i.e., especially proteins above ≈75 kDa), the *in gel* or electroelution approach may be superior.

In this chapter generalized procedures are given for enzymatically digesting proteins in solution as well as *in situ* in SDS–PAGE gels and for eluting PVDF-blotted proteins in a form that is amenable to subsequent enzymatic digestion. Particular attention is given to highlighting those parameters that are most important in terms of ensuring the overall success of these approaches. Finally, a short description is provided of reverse-phase HPLC fractionation of subnanomole amounts of enzymatic digests.

SAMPLE PREPARATION FOR ENZYMATIC DIGESTION

Preparation of Non-SDS Containing Samples

The first (and in many respects the most important) step in enzymatically cleaving a protein is preparing the sample in a form that is suitable for enzymatic digestion. For proteins not isolated from an SDS gel, one of the pathways in the flow chart shown in Fig. 1 can usually be followed. If the sample contains less than the equivalent of about 0.1 mmol monovalent salt, then it

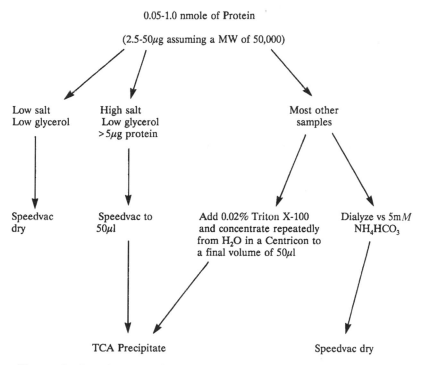

Figure 1   Sample preparation schematic for non-SDS-containing samples. Protein samples that are in low salt and glycerol can be dried in a Speed-Vac before proteolysis. Other samples may be concentrated by precipitation or by centrifuging in a Centricon filter (after the addition of Triton X-100 to prevent nonspecific adsorption, note that although this detergent co-precipitates with the protein following the addition of TCA, it can then be readily extracted with acetone). Samples in large volumes of nonvolatile salts and buffers may have to be dialyzed against a volatile buffer prior to concentrating.

can simply be dried in a Speed-Vac (note that the final digest can be done in up to ≈ 1 *M* NaCl). If the glycerol concentration is low but the amount of salt is excessive, then one option is to precipitate the protein with TCA. In either instance, the final precipitation or drying should be carried out in the same 1.5-ml Eppendorf tube that is used for the enzymatic digestion.

TCA Precipitation Procedure

TCA precipitation is effective at removing salts and many detergents (such as SDS) from the protein prior to enzymatic digestion.

1. Concentrate the sample to a protein concentration >100 µg/ml.

2. Add one-ninth the total volume of the sample of 100% TCA for a final TCA concentration of 10%.

3. Incubate on ice for 30 min.

4. Microfuge and remove the supernatant. Save the supernatant in case the protein did not precipitate.

5. Wash the protein pellet twice with 100 µl of cold acetone.

6. Let the sample air dry.

TCA Precipitation Troubleshooting

Efficient precipitation requires that the protein concentration be as high as possible and that the glycerol concentration be less than 15–20% (v/v). Typically, during TCA precipitation the protein concentration should be at least 100 µg/ml. In many cases the protein can simply be concentrated in a Speed-Vac to increase the concentration to this level. If the amount of protein precipitated is less than a nanomole, a visible pellet may not be observed.

Preparation of SDS Gel-Eluted Proteins

Proteins that have been isolated from SDS–PAGE gels must be treated in some way so that the final SDS concentration in the digest is less than 0.0005%. Under the conditions that we recommend (i.e., when the digest is carried out in the presence of 2 *M* urea), SDS concentrations above 0.05% completely prevent trypsin and endoproteinase Lys-C digestion (Fig. 2). Similar amounts of SDS (i.e., 200 μl of 0.05%) also interfere with reverse phase HPLC. When amounts of SDS above 0.1 mg are

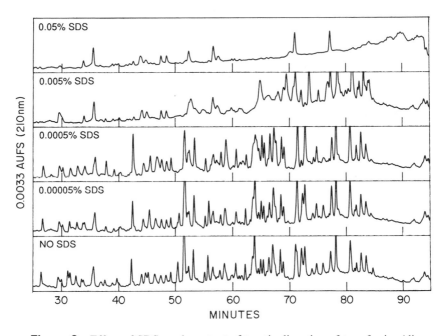

Figure 2 Effect of SDS on the extent of tryptic digestion of transferrin. Aliquots of transferrin (50 pmol) were added to various concentrations of SDS and then digested with trypsin as described in the text. Digests were then subjected to reverse-phase HPLC on a Vydac C-18 2.1 × 250-mm column and the resulting peptides were eluted with increasing concentrations of acetonitrile using a flow rate of 0.15 ml/min. The bottom chromatogram is a 50 pmol control digest that does not contain SDS. Concentrations of SDS above 0.0005% (in a volume of 200 μl) cause a significant decrease in the number of peptides resolved. At 0.05% SDS, essentially no digestion was observed. (From K.L. Stone, M.B. LoPresti, and K.R. Williams, "Enzymatic Digestion of Proteins and HPLC Peptide Isolation in the Subnanomole Range", *in* Laboratory Methodology in Biochemistry, 1990. Reprinted with permission of CRC Press, Inc.)

injected onto a reverse-phase HPLC column there is a general decrease in resolution and later elution of peptides *(60)*. Although techniques such as "inverse gradient" reverse phase HPLC *(59)* and guanidinium–HCl precipitation *(48)* can be used to remove SDS from the protein, we generally prefer acetone precipitation which provides a simple and reliable means of reducing the SDS concentration to a level that no longer interferes with enzymatic digestion.

Acetone Precipitation Procedure

1. Extensively dialyze 1 ml or less of eluted protein versus 0.05% SDS, 5 mM NH$_4$HCO$_3$.

2. Transfer to a 1.5 ml Eppendorf tube and Speed-Vac dry.

3. Add 50 µl of water.

4. Add 450 µl of cold acetone, 1 mM HCl.

5. Incubate at −20°C for 3 hr.

6. Microfuge and remove the supernatant. Save the supernatant in case the protein did not precipitate.

7. Wash the "pellet" twice with 100 µl cold acetone.

8. Air dry the pellet.

Acetone Precipitation Troubleshooting

As with TCA precipitation, the protein concentration should be kept above 100 µg/ml. Amounts of SDS greater than the equivalent of approximately 1 ml of 0.05% SDS will tend to co-precipitate with the protein forming a large white precipitate upon the addition of the cold acetone. If this occurs, the dried sample should be resuspended in 50 µl H$_2$O and steps No. 4–8 repeated.

Proteins Isolated by Electroblotting onto PVDF Membranes

The use of SDS-polyacrylamide gel electrophoresis followed by electroblotting onto PVDF membranes *(84)* has become a primary method for the purification of proteins. PVDF membranes have high specificity for binding proteins compared to salts, amino acids, or detergents so the resulting protein is isolated free from these contaminants. The method below for eluting PVDF blotted proteins is an extension of the Yuen *et al. (89)* procedure which takes advantage of the relatively weak affinity of Immobilon-P membranes for low-molecular-weight proteins by cleaving the protein *in situ* with CnBr prior to eluting with a volatile solvent.

Procedure for CnBr Eluting PVDF-Blotted Proteins

1. Cut the Immobilon-P PVDF membrane containing the blotted protein band into approximately 1-mm-wide pieces and place in a 1.5-ml Eppendorf tube. Cut an equal size area of membrane from a section of the blot which does not contain protein to be used as a control to identify artifact and other reagent peaks in the final HPLC chromatogram.

2. Add 500 µl of ice cold 95% acetone to the membrane pieces.

3. Vortex and then shake gently on a rocker platform at 4° C for 1 hr or more.

4. Remove the acetone supernatant and save.

5. Add 500 µl of ice-cold 95% acetone and vortex. Combine supernatant with previous supernatant.

6. Air dry the PVDF pieces.

7. Suspend the PVDF pieces in 100 μl of 70% formic acid.

8. Make up a 70 mg/ml or 0.66 M CnBr stock solution: 132 μl of 5 M CnBr in acetonitrile and 868 μl of 70% formic acid.

9. Add 10 μl of the 70 mg/ml CnBr solution per 10 μg of blotted protein.

10. Incubate at room temperature for 24 hr in the dark.

11. After 24 hr, transfer the supernatant (which contains a significant fraction of the CnBr fragments) to another 1.5-ml Eppendorf tube and Speed-Vac to dryness. Also, Speed-Vac the original tube containing the PVDF pieces to remove residual formic acid.

12. Add 100 μl of 40% acetonitrile to the tube containing the PVDF pieces and then incubate at 37°C for 3 hr.

13. Remove the supernatant and combine with the dried CnBr/ formic acid supernatant.

14. Add 100 μl of 40% acetonitrile/0.05% TFA to the PVDF pieces and then incubate at 50°C for 20 min.

15. Combine this supernatant with the CnBr supernatant from step No. 11 and Speed-Vac to dryness.

16. Add 60 μl of water and re-dry.

CnBr Elution Troubleshooting

Dyes. Prior to *in situ* CnBr cleavage, it is important to minimize the amount of dye on the PVDF membrane. If the blot was stained with Coomassie blue, then the above acetone washing procedure should be used. If Ponceau S was used, then steps No. 2–5 can be replaced with two brief washings with 500 μl H_2O. The problem with dyes, especially Coomassie blue, is that high concentrations of Coomassie blue inhibit enzymatic digestion and lead to artifact peaks that elute at about 70% acetonitrile in the reverse-phase HPLC chromatogram.

Efficiency of CnBr Elution/Choice of Membrane. The efficiency of elution from Immobilon-P membranes of 11 standard proteins ranging in molecular mass from 14 to 116 kDa, was 74.1% *(106)*. This technique, however, is not compatible with Problott membranes because these membranes have a higher affinity than Immobilon-P membranes for low-molecular-weight proteins (i.e., our studies suggest that while both membranes have similar blotting efficiencies for proteins that are larger than about 25 kDa, ProBlott generally has higher blotting efficiencies for proteins that are less than about 25 kDa). The relatively low affinity of Immobilon-P membranes for low molecular weight proteins (i.e., CnBr peptides) makes it an ideal matrix for CnBr elution. Hence, the average elution efficiency of four standard proteins that ranged from 25 to 75 kDa, was 17% from ProBlott as compared to 76% for Immobilon-P membranes.

Surface Area of PVDF. Optimum results require that the protein/PVDF surface area ratio is kept as high as possible with the best results being obtained when this ratio is above $\approx 2 \ \mu g/cm^2$. As this ratio is decreased fivefold, the efficiency of CnBr elution decreases by as much as 50%.

Proteins Isolated by SDS–PAGE for in-Gel Digestion

An alternative approach for obtaining internal sequences from SDS–PAGE-separated proteins is to carry out the digestion in the gel *(59, 113, 113a)*. This approach eliminates the electrophoretic transfer to PVDF membranes and thereby, the losses that occur during this step. Below is a modified gel staining and destaining procedure that should be used to prepare the sample for in gel enzymatic digestion. The SDS–PAGE gel can range from 7 to 15% acrylamide. Because of the difficulty in blotting some large proteins (i.e., over ≈75 kDa), in these instances peptide yields may be somewhat increased by using the in gel as opposed to the PVDF blotting/elution approach.

SDS–PAGE Staining and Destaining Procedure for Use with the in-Gel Digestion Procedure

1. Stain the gel with 0.1% Coomassie blue in 10% acetic acid, 50% methanol, and 40% H_2O for 60 min.

2. Destain the gel by soaking for 2 hr in 10% acetic acid, 50% methanol, and 40% H_2O. The gel may still have a relatively dark Coomassie Blue background.

3. Excise the band from the gel in such a manner as to avoid removing gel that does not contain protein. Also remove a "blank" piece of gel which will be used as a control.

4. Place the excised band in an Eppendorf tube.

5. The sample is now ready for in-gel digestion.

ENZYMATIC DIGESTION OF PROTEINS

Once the protein has been prepared using one of the above approaches, it can then be cleaved using any one of several different proteases. Trypsin cleaves specifically at the carboxy-terminal side of lysine and arginine (excluding lysine–proline and arginine–proline linkages). Endoproteinase Lys-C cleaves specifically after lysines while chymotrypsin cleaves primarily on the carboxy-terminal side of aromatic amino acids with more limited cleavage at leucine, methionine, and other hydrophobic amino acid side chains. In our experience, trypsin and chymotrypsin, but not endoproteinase Lys-C, will successfully digest proteins that are not soluble in the 2 M urea, 0.1 M NH$_4$HCO$_3$ used for carrying out the digestion. Typically, in these instances the solution will clear within an hour of adding the trypsin or chymotrypsin. Trypsin is commonly preferred for obtaining complete enzymatic digestion because of its specificity and ability to digest insoluble and adsorbed proteins.

Trypsin, Chymotrypsin, and Endoproteinase Lys-C Digestion Procedure

Samples that have been either precipitated or CnBr eluted from Immobilon-P membranes or simply dried are now ready for enzymatic digestion as described below.

1. Add 25 µl 8 M urea, 0.4 M NH$_4$HCO$_3$ to the dry protein in a 1.5-ml Eppendorf tube.

2. Check the pH by spotting 1–2 µl on pH paper. The pH should be between 7.5 and 8.5.

3. Remove 5–15% for acid hydrolysis followed by ion exchange amino acid analysis.

4. Add 5 µl of 45 mM dithiothreitol.

5. Incubate at 50°C for 15 min.

6. Add 5 µl of 100 mM iodoacetamide after cooling to room temperature.

7. Incubate at room temperature for 15 min.

8. Add 60 µl water.

9. Add a 1:25, enzyme to protein, weight-to-weight ratio of trypsin, chymotrypsin, or endoproteinase Lys-C in a volume of 5 µl.

10. Incubate at 37°C for 24 hr.

11. Stop reaction by freezing or by injecting directly onto a reverse-phase HPLC column.

Trypsin, Chymotrypsin, and Endoproteinase Lys-C Digestion Troubleshooting

Amount of Protein. One of the major reasons for digests not working is simply that the amount of protein has been considerably overestimated because of losses incurred in final sample preparation or to the inaccuracy of colorimetric and dye-binding assays. For this reason, it is essential to obtain an accurate amino acid analysis either immediately before drying the sample in the Eppendorf tube (which will be used for the digest) or after redissolving the protein in the 25 µl 8 M urea, 0.4 M NH$_4$HCO$_3$. While 10 µl 8 M urea, 0.4 M NH$_4$HCO$_3$ buffer is compatible with ion-exchange amino acid analysis (when used solely for the purpose of determining protein concentrations), this amount of urea is not well tolerated by PTC amino acid analysis. In the latter case, an aliquot of the protein should either be taken for amino acid analysis prior to drying it down (in preparation for dissolving in 25 µl 8 M urea) or the dried protein should be dissolved in 70% formic acid or 100% trifluoroacetic acid. After removing an aliquot for amino acid analysis, the remaining solution can then simply be evaporated in a Speed-Vac prior to dissolving in urea. Although we have succeeded in sequencing tryptic peptides obtained from as little as 25 pmol of protein, we generally recommend that the digest not be attempted unless the amino acid analysis indicates that a minimum of 100 pmol protein is available.

Protein Concentration. The minimum final protein concentration (in the 100 µl 2 M urea, 0.1 M NH$_4$HCO$_3$) that is required in order to obtain a reasonably complete digestion is 20 µg/ml

(58). If the concentration is significantly below this level, the extent of digestion decreases substantially.

Buffers, Salts, and Detergents. Enzymatic digestion can be inhibited or completely prevented by various reagents that may be present in the sample. As mentioned previously, in combination with the 2 M urea, 0.1 M NH_4HCO_3, as little as 0.0005% SDS will effect the extent of enzymatic digestion. SDS concentrations above 0.005% will substantially reduce the yield of many tryptic peptides. Similarly, high levels of Coomassie blue and ampholines will also prevent digestion. In our experience, even prolonged dialysis extending over several days is not sufficient to reduce ampholine concentrations to a level that permits trypsin digestion using the above protocol. Rather, a technique such as TCA precipitation or hydrophobic chromatography is required to free the protein of ampholines. Relatively high levels of monovalent salts such as NaCl can, however, be tolerated in the final digest (i.e., 1 M NaCl has relatively little effect) *(58)*. Following carboxamidomethylation, the urea concentration must be reduced to 2 M by diluting with water prior to adding the enzyme. While 2 M urea is sufficient to keep most proteins in solution, it is below the level (approximately $3M$) that significantly inhibits trypsin digestion. Extensive testing with 50-pmol amounts of several peptides demonstrates that, under the conditions recommended above, no detectable NH_2-terminal blocking arises from cyanate formation in the urea *(58)*. Stock solutions of 8 M urea, 0.4 M NH_4HCO_3 that are made from ultrapure urea may be stored frozen for up to 2 weeks prior to use.

Carboxamidomethylation. As judged by the yield of the resulting tryptic peptides, dissolving a protein in 8 M urea, 0.4 M NH_4HCO_3 and heating does not appear to bring about complete and irreversible denaturation. For this reason, prior carboxamidomethylation, which irreversibly modifies all cysteine residues, usually brings about a marked improvement in the extent of digestion. The value of this approach is clearly shown in

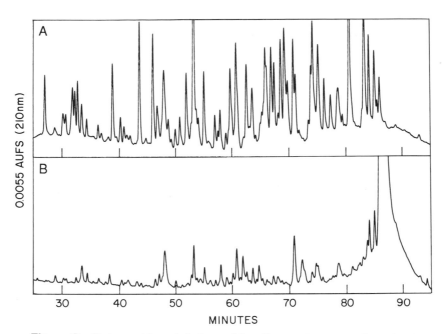

Figure 3 Carboxamidomethylation and its effect on the extent of trypsin digestion. In each chromatogram, 50 pmol of transferrin was digested with trypsin. (A) Transferrin was carboxamidomethylated prior to digestion while (B) was not. Peptides were isolated on a 2.1-mm × 25-cm Vydac C-18 column that was eluted at a flow rate of 0.15 ml/min as described in the text. (From K.L. Stone, M.B. LoPresti, and K.R. Williams, "Enzymatic Digestion of Proteins and HPLC Peptide Isolation in the Subnanomole Range", *in* Laboratory Methodology in Biochemistry, 1990. Reprinted with permission of CRC Press Inc.)

Fig. 3, where 50 pmol aliquots of transferrin were digested with trypsin either with (Fig. 3A) or without (Fig. 3B) prior carboxamidomethylation. Under the conditions used above, the excess carboxamidomethylation reagents do not interfere with the subsequent trypsin digestion or HPLC purification of the resulting peptides, thus eliminating the necessity of dialyzing or gel filtering the denatured protein. The latter is frequently associated with a large loss of protein.

Amount of Enzyme. An enzyme to protein (w/w) ratio of 1:25 is sufficient to adequately digest the protein while keeping the amount of trypsin to a minimum *(58)*. For proteins that are larger than 60 kDa, this ratio should be decreased to 1:50 so that the mole ratio of substrate/trypsin protein is at least 10. At this mole ratio, peptides resulting from trypsin autolysis are seldom a problem. Even so, this possibility can be guarded against by carrying out an enzyme control with the same amount of trypsin (but no substrate protein) and by immediately searching all of the resulting peptide sequences against the Protein Identification Resource, Genbank and Swiss-Protein databases (to identify previously sequenced proteins and to absolutely rule out the possibility of the sequence arising from trypsin).

Trypsin (as well as endoproteinase Lys-C) can be purchased from Boehringer-Mannheim. It is important to use an appropriate preparation of trypsin (such as "sequencing grade") since impure trypsin will decrease the yield of individual peptides and cause an increase in the complexity of the HPLC profile *(54)*. Stock solutions (1 mg/ml) of trypsin in 1 mM HCl are stable for at least 6 months if stored in individual aliquots at −20°C.

In Situ Enzymatic Digestion in SDS–PAGE Gels

Proteins that are purified by SDS–PAGE can be enzymatically digested directly in the gel matrix. This technique appears to be especially well suited for large proteins (>75 kDa). Prior to in gel digestion, the amount of protein can be quantitated by hydrolysis of a 10–15% slice of the gel followed by ion–exchange amino acid analysis.

1. For Coomassie blue stained samples, add 500 µl of ice cold 95% acetone to the gel pieces containing the protein.

2. Shake at 4°C for 30 min.

3. Remove supernatant and save.

4. Speed-Vac dry the gel pieces.

5. Add 500 µl of 0.1 M NH_4HCO_3 to the gel sample and to an equal size "blank" piece of gel. The latter will be used as a control to identify reagent and other artifact peaks in the final HPLC chromatogram.

6. Shake (on a rocker platform) at room temperature for 4 hr or more.

7. Remove wash and save.

8. Add another 500 µl 0.1 M NH_4HCO_3 to the gel pieces and shake overnight (16 hr).

9. Remove wash and save.

10. Add 150 µl of 0.1 M NH_4HCO_3 to the gel pieces, if the gel absorbs all of the liquid then this volume may be increased appropriately.

11. Add 5 µl of 45 mM dithiothreitol to the gel pieces.

12. Incubate at 50°C for 20 min.

13. Remove samples from the incubator; cool to room temperature, and add 5 µl of 100 mM iodoacetamide.

14. Incubate at room temperature in the dark for 20 min.

15. Add trypsin in a 1:25 weight-to-weight ratio of enzyme to protein.

16. Incubate at 37°C for 24 hr. During this time most or all of the liquid may have been absorbed by the gel.

17. Add 500 µl of 0.1 M NH$_4$HCO$_3$ to the gel pieces and shake at room temperature for 8 hr to extract the peptides.

18. After 8 hr, transfer the supernatant to a 1.5-ml Eppendorf tube.

19. Add 300 µl of 2 M urea, 0.1 M NH$_4$HCO$_3$ to the gel pieces.

20. Shake at room temperature for 24 hr.

21. After 24 hr, transfer the supernatant to the corresponding Eppendorf tube.

22. Speed-Vac dry the combined supernatants.

23. Redissolve the supernatant in 210 µl of H$_2$O.

24. Transfer the supernatant to a 0.22-µm UltraFree Millipore filter unit and microfuge.

25. The sample may now be injected onto a reverse-phase HPLC column.

In Situ Digestion Troubleshooting

Minimization of SDS. The SDS is brought to sufficiently low levels by staining/destaining the gel as well as by the extensive washing procedure described in steps No. 1–9 above. If enough SDS is not removed prior to reduction, alkylation and enzymatic digestion, the protein will not digest efficiently *(56, 60)* and, in addition, problems will be encountered with reverse-phase HPLC *(60)*.

Amount of Gel. The amount of gel (at least up to an approximately 0.5 × 12-cm slice) does not appear to be extremely critical although it is obviously best to maximize the protein/volume of gel matrix ratio. It is also considerably more difficult to work with a larger volume of gel.

Kathryn L. Stone and Kenneth R. Williams

HPLC FRACTIONATION OF SUBNANOMOLE AMOUNTS OF ENZYMATIC DIGESTS

The following provides a summary of general conditions that can be used to isolate peptides from enzymatic digests. A more detailed summary, which includes a discussion of important parameters to consider when selecting appropriate HPLC systems, columns and running conditions, can be found in Ref. *(57)*.

Buffer A = 0.06% TFA, H_2O
Buffer B = 0.052% TFA, 80% acetonitrile
Injection volume = 200 µl
UV detection = 210 nm

Gradient:

0–60 min	2.0–37.5% B	
60–90 min	37.5–75.0% B	
90–105 min	75.0–98.0% B	

Analytical HPLC conditions

Amount of sample: 0.25–20.0 nmol

Columns: 4.6 mm × 25 cm
Vydac C-18 reverse phase
300 Å , 5 µm support
3.9 mm × 15 cm
Delta Pak C-18 reverse phase
300 Å , 5 µm support
Flow rate: 0.5 ml/min

Narrowbore HPLC conditions

Amount of sample: 50–250 pmol

Columns: 2.1 mm × 25 cm
Vydac C-18 reverse phase
300 Å , 5 µm support
2.0 mm × 15 cm
Delta Pak C-18 reverse phase
300 Å , 5 µm support

Flow Rate: 0.15 ml/min

Notes on Reverse-Phase HPLC

Mobile Phase

The pH 2.2, 0.05% TFA/acetonitrile buffer has become an almost universal mobile phase for reverse-phase isolation of peptides. Its low UV absorbance, volatility, and good solubilizing and resolving properties make it the current mobile phase of choice. The concentration of TFA in Buffer A is slightly higher than in Buffer B to help balance the absorbance of the buffers *(57)*. If the absorbance trace slopes upwards, additional TFA (i.e., typically 50–150 µl of a 20% (v/v) stock/liter of mobile phase) can be added to Buffer A to effectively balance the baseline. Conversely, if the baseline slopes downward, the TFA concentration can be increased in Buffer B.

Reverse-Phase Columns

In general, the longest and narrowest columns should be used to optimize resolving power and sensitivity, respectively. Figure 4 demonstrates the loss in resolution that occurs as the length of the column is decreased. Hence, the number of peptide peaks detected decreased from 120 on the 25-cm column to 90 on the 15-cm column and to 77 on the 5-cm column. The number of peaks detected on the 5-cm column is thus 35% less than on the 25-cm column. Although Delta Pak columns are not available in 25-cm lengths, the apparently greater resolving power of this support enables a 15-cm Delta Pak column to give a similar level of resolution as that obtained on a 25-cm Vydac column. The selectivity of these two supports also appears to be very similar *(57)*. Since we have only tested a limited number of reverse-phase columns it is quite possible that other supports give equal or even better resolution than the two columns that are recommended above. In our experience, however, the best results are obtained on 300 Å, 5 µm supports. We have also found that col-

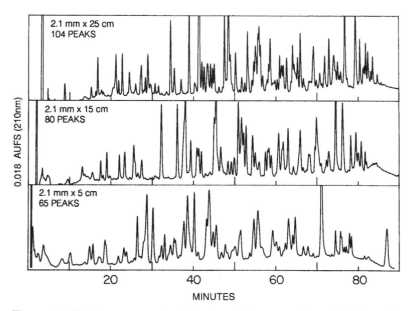

Figure 4   The importance of column length in reverse-phase HPLC peptide separations. Tryptic peptides from 50 pmol of carboxamidomethylated transferrin were injected onto each of the three indicated Vydac C-18 reverse-phase columns and eluted at 0.2 ml/min as described in the text. The column dimensions and number of peaks detected are indicated. (From K.L. Stone, M.B. Lo-Presti, and K.R. Williams, "Enzymatic Digestion of Proteins and HPLC Peptide Isolation in the Sub-nanomole Range", *in* Laboratory Methodology in Biochemistry, 1990. Reprinted with permission of CRC Press, Inc.)

umn plate counts determined with small organic molecules are extremely poor predictors of a reverse phase column's ability to resolve complex enzymatic digests. When comparing columns obtained from the same manufacturer, we have previously *(52)* noted very little change in selectivity as the support is changed from a C-4 to a C-8 or C-18 chain length. The major differences being that as the chain length is increased, the peak retention time is also increased and some peptides that elute early on the C-18 support were not retained on the C-4 column. Because of the latter finding we routinely use a C-18 column.

The choice between a "conventional" 3.9- to 4.6-mm I.D. versus a 2.0- to 2.1-mm I.D. narrowbore column is determined primarily by the amount of digest that is being fractionated. Our previous studies *(60)* suggest that in order to obtain optimum resolution, amounts of enzymatic digests above 250 pmol should be chromatographed on conventional 3.9- to 4.6-I.D. columns.

Flow Rates

The major advantage in using narrowbore columns is that the flow rates can be decreased from the usual 0.5–1 ml/min used for conventional columns to the ≈150-μl/min range which increases the sensitivity of detection and reduces peak volume. The latter may help avoid having to concentrate the peptides prior to sequencing. This is an important consideration as large losses are sometimes incurred when concentrating subnanomole amounts of peptides. In practice, the lowest flow rate consistent with near maximum resolution should be used. As long as the gradient times are maintained constant, our studies suggest that there is essentially no loss in resolution as the flow rate is decreased from 1.0 to 0.4 ml/min on a 4.6 mm I.D. column. As shown in Fig. 5, the number of peaks detected at 0.4 ml/min is 92% of that observed at the optimum flow rate of 0.8 ml/min. Hence, we find that a flow rate of 0.5 ml/min on an analytical column represents a reasonable compromise between peak resolution and peak volume. In the case of narrowbore columns, we have encountered problems in using automatic peak detectors below a flow rate of 150 μl/min (i.e., at flow rates much below 150 μl/min more than one peak frequently elutes during the time the peak detector is counting down the transit time from the UV detector to the fraction collecto; hence, the peaks are mixed together). Therefore, we routinely use a flow rate of 150 μl/min on 2.0- to 2.1-I.D. columns.

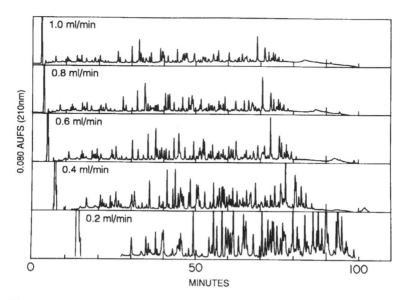

Figure 5 The effect of flow rate on the HPLC separation of tryptic peptides. In each chromatogram 250 pmol of a tryptic digest of carboxamidomethylated transferrin was injected onto a 4.6-mm × 25-cm Vydac C-18 reverse-phase HPLC column. The column was eluted at the flow rates indicated as described in the text. The number of peaks detected were 112 at 1.0 ml/min, 117 at 0.8 ml/ min, 112 at 0.6 ml/min, 108 at 0.4 ml/min and 96 at 0.2 ml/min. (From K.L. Stone, M.B. LoPresti, and K.R. Williams, "Enzymatic Digestion of Proteins and HPLC Peptide Isolation in the SubNanomole Range", *in* Laboratory Methodology in Biochemistry, 1990. Reprinted with permission of CRC Press, Inc.)

Repurification of Peptides

Our previous studies *(57)* have demonstrated a significant difference in selectivity between an Aquapore C-8 versus the Vydac or Delta Pak C-18 columns. As a result of this observation, which does not appear to be due to the different chain lengths (see above), we routinely "screen" small aliquots of peptides isolated on a Vydac C-18 column on a 1.0 × 250 mm Aquapore column operated at 150 μl/min. In our experience, the latter column gives better resolution than the 2 mm Aquapore cartridge. Those peptides that are found to be impure are then diluted with an equal volume of 2 M urea and repurified by injecting onto the Aquapore column using the same gradients as listed above. An alternative means of repurifying would be to change the mobile phase to a pH 6 phosphate system which gives approximately 85% of the peak resolution obtainable with the TFA system *(57)*. However, the latter mobile phase is not volatile and, if only one HPLC system is available, it is usually much easier to change the column rather than the mobile phase.

SUMMARY OF RESULTS FROM THE CYANOGEN BROMIDE/ACETONITRILE ELUTION AND TRYPTIC DIGESTION OF 26 "UNKNOWN" PROTEINS

As shown in Fig. 6, similar tryptic peptide maps were obtained for 65 pmol amounts of transferrin that either had just been digested in solution or had first been CnBr eluted from a PVDF membrane. It is apparent that a similar extent of cleavage occurred in each case and that few, if any, extraneous peaks are present in the PVDF-eluted sample (which had also been stained with Coomassie blue). A similar comparison of the in gel digest protocol is shown in Fig. 7. Again, the profiles obtained from tryptic digests carried out in solution are similar to those carried out *in situ* in the gel. To better assess the capabilities of the cyanogen bromide/acetonitrile elution approach to internal amino acid sequence analysis, we compiled a summary of the results that were obtained on 26 proteins that were recently submitted for analysis *(106)*. As judged by the fraction of these proteins for which tryptic peptide sequences were obtained that were

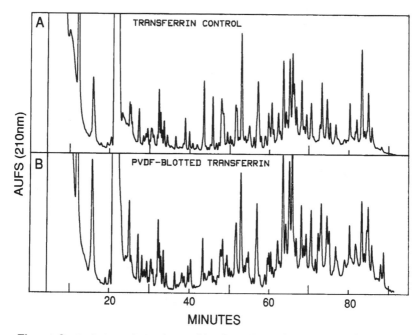

Figure 6 Isolation of tryptic peptides from 5 μg (65 pmol) transferrin that had either been digested in solution or cyanogen bromide eluted from Immobilon-P membranes. In B, 15 μg of transferrin was subjected to SDS–PAGE and then electroblotted with an efficiency of about 32% onto Immobilon-P. The membrane was then CnBr eluted and the resulting peptides were digested with trypsin as described in the text. (A) A nonblotted control that was cleaved with cyanogen bromide and digested with trypsin in solution. It is apparent that a similar extent of cleavage occurred in each of these samples and that there are few if any extraneous peaks present in the PVDF-eluted sample that had also been stained with Coomassie blue.

suitable for generating cDNA probes, the overall success rate was 96%. Four of these proteins had been stained with Ponceau S while the remainder had been stained with Coomassie blue. Since nine of these samples contained between 36 and 71 pmol protein, it is possible to routinely succeed with approximately 50 pmol eluted protein. Approximately 70% of the peaks chosen

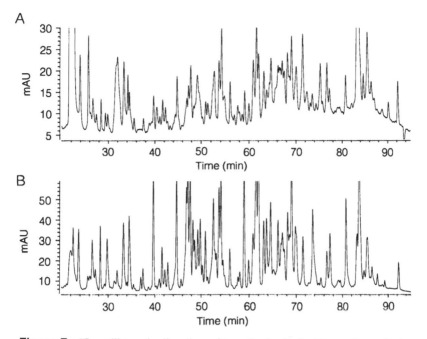

Figure 7 "In-gel" tryptic digestion of transferrin. In A, 15 μg of transferrin was subjected to SDS–PAGE and digested in the gel matrix using the procedure described in the text. (B) A control transferrin digest on 15 μg that was not run on SDS–PAGE. Based on amino acid analysis of the eluted peptides from the in-gel digestion, 47.6% of the protein was recovered from the in gel sample. These chromatograms have been normalized for this difference in amount.

for sequencing provided unambiguous sequence. By giving preference to "later" eluting peaks, it was possible to achieve an average peptide length (in terms of positively called residues) of about 12 residues. In many instances the latter corresponded to the complete sequence of the peptide. For comparison, based on the average abundance of arginine (5.4%), lysine (5.7%), and methionine (2.3%) in proteins in the Protein Identification Resource Database, the average length of a peptide expected from this procedure would be seven to eight residues.

3

Purification of Proteins and Peptides by SDS–PAGE

Nancy LeGendre,* Michael Mansfield,* Alan Weiss,* and Paul Matsudaira†

*Millipore Corporation
Bedford, Massachusetts

†Whitehead Institute for Biomedical Research and Department of Biology
Massachusetts Institute of Technology
Cambridge, Massachusetts

Gel Electrophoresis
Electroblotting to PVDF Membranes
  Choice of PVDF Membrane
  Electrotransfer
  Visualization of Proteins
  Yields
Removal of N-Terminal Blocking Groups from
  Electroblotted Proteins or Peptides
  Removal of Blocking Groups from N-Acetylserine and
    N-Acetylthreonine Residues
  Removal of Formyl Groups from N-Terminal Met Residues
  Removal of Pyroglutamate from N-Terminal Residues
  Removal of Blocking Groups from N-Acetylamino Acid
    Residues
Electroelution from SDS Gels
Amino Acid Analysis of Electroblotted Proteins
Phosphoamino Acid Analysis of Electroblotted
  Proteins
Desorption of Electroblotted Proteins from PVDF
  Membranes

NOTES

Current gas-phase sequencing instruments require that the protein sample be in a small volume (<100 μl) and in a volatile solvent (H_2O, acetonitrile, methanol, or dilute acetic or formic acid). Although HPLC-purified samples are sufficiently pure and clean to satisfy both requirements, HPLC is often the rate-limiting step in obtaining samples. In addition, a well-resolved peak from an HPLC column may still be composed of several polypeptides. A simple alternative method for purifying proteins is SDS–PAGE.

To avoid interference with sequencing chemistry (see Introduction), proteins separated by electrophoresis must be eluted from the gel and purified from the electrophoresis reagents and buffers. Protein in the gel can be visualized by staining with a dye such as Coomassie blue. The stained band is cut from the gel and electroeluted (77). The protein is recovered from the electroeluter but must be concentrated and separated from contaminants by precipitation, HPLC, or dialysis. Although electroelution is very efficient for recovering protein, subsequent steps reduce yields. To improve the recovery of proteins from gels, electroblotting has become an attractive alternative. Methods have been developed for electroeluting proteins onto derivatized glass-fiber filters (65, 83, 85), although the protein cannot be visualized directly by charged dyes and must be chemically modified prior to electrophoresis. One improvement of the electroblotting method (82, 84, 87) is to transfer the protein onto PVDF membranes. PVDF can be cast as a microporous thin film that is not damaged by the reagents and solvents used in commercial, automated sequencing instruments. Proteins bind to the membrane through hydrophobic interactions and can be visualized by staining, while contaminants are washed from the membrane by rinsing with water.

GEL ELECTROPHORESIS

Reagents

Electrophoresis-grade or higher quality reagents should be used for gels and buffers. Alternatively, precast gels may be used. Solutions should be dissolved in Milli-Q (Millipore) grade water.

Procedure

Samples can be separated by one- or two-dimensional gel electrophoresis [e.g., Refs. *(62, 64)*]. Approximately 25–100 pmol of protein, usually corresponding to 1–5 µg, should be loaded. Samples separated on 8 × 10-cm minigels (0.5 mm thick) with narrow teeth produce bands of ideal size (2 × 4 mm). Larger bands might not fit in the sequencer's reaction cartridge. Dilute samples should be either concentrated and loaded in one lane of a minigel or loaded directly into a single lane of a thicker (1–3 mm) gel. β-Lactoglobulin or bovine serum albumin loaded in adjacent lanes can serve as an internal control to measure the efficiency of transfer and to detect N-terminal blockage.

Proteins can also be sequenced from blots of two-dimensional gels. Ampholytes and urea do not interfere with sequence analysis because they can be washed from the blot. Proteins can be sequenced from blots of isoelectric focusing gels cast using immobilized pH gradients (Immobiline, LKB). The major difference in obtaining protein from a two-dimensional gel is that the amount of sample that can be loaded on an isoelectric focusing gel (the first dimension) is less than on a corresponding SDS one-dimensional gel. Thus, spots from 5 to 10 two-dimensional gel blots are usually pooled and sequenced simultaneously.

If the protein band (spot) is well separated from neighboring bands, it can be cut out with a razor blade. Generally, any protein that can be resolved by electrophoresis can be sequenced from a blot. However, large proteins (>100 kDa) may prove difficult to transfer to membranes quantitatively. Small peptides (<8–10 kDa) may fail to bind to the membrane during transfer or may wash off the membrane during the sequencing cycles. Peptide wash-off is minimized by treating the membrane with polybrene either before or after electroblotting.

Blockage of the N terminus during electrophoresis is thought to be a major factor in reducing the sequenceable amount of protein. Moos *et al. (87)* have described a gel system to reduce N-terminal modification during electrophoresis. Generally, if high quality reagents and water are used, gels can be run without pre-electrophoresis or cooling. If blockage is a problem, the gel can be subjected to pre-electrophoresis immediately prior to sample loading *(5, 15)*.

ELECTROBLOTTING TO PVDF MEMBRANES

Choice of PVDF Membrane

In 1987, utilization of PVDF (Immobilon-P, Millipore) as a substrate for protein sequencing *(84)* was first described. Subsequently, additional PVDF membranes have been introduced as alternative substrates. While all of these membranes are functional in sequencing, their performance varies in different applications. The choice of membrane should be made depending on the sequencing strategy used.

Based on performance criteria, the PVDF membranes currently offered can be categorized into two groups. The first group includes Immobilon-P and Westran (Schleicher and Schuell, Inc.). While Immobilon-P is composed of pure PVDF, Westran is a composite membrane consisting of PVDF on a polyester web support. Both membranes have nominal pore sizes of 0.45 μm. The structures of these membranes permit a comparatively high level of protein passage during electroblotting. Consequently, valuable protein samples can be lost to the transfer buffer. Generally, the degree of protein passage increases as the size of the protein decreases. One method for retaining more of the protein is to include a backup sheet of membrane. These membranes perform comparably in most blotting applications; however, composite membranes may deteriorate during sequencing.

The second group of membranes includes Immobilon-P^{SQ} (Millipore), ProBlott (Applied Biosystems, Inc.), Trans-Blot PVDF (Bio-Rad), and Fluorotrans (Pall Corporation). Like Immobilon-P, these membranes are composed of pure PVDF. Their nominal pore sizes, however, are all about 0.1 μm. Because the binding capacities of these membranes are higher, protein passage during electroblotting is greatly reduced. Under standard transfer conditions, these membranes can bind up to 100% of proteins of $M_r > 15-20$ kDa. Colloidal gold staining of backup membranes does not detect any proteins. Protein passage is also greatly reduced for proteins of $M_r < 15$ kDa. The ability to capture more protein during electrotransfer is reflected in increased initial sequencing yields with comparable repetitive yields. While the membranes are excellent substrates when only N-terminal sequence is required, the tenacity with which they bind protein makes them unsuitable when the bound protein must be subjected to chemical or proteolytic cleavage for internal sequence analysis. An appropriate strategy, described in Chapters 2 and 4, is to transfer the protein to Immobilon-P, perform the cleavage on the membrane, and then resolve proteolytic fragments by reverse-phase HPLC.

Electrotransfer

Two systems exist for efficient electrotransfer of proteins to PVDF membranes. Tank transfer *(65, 84)* requires immersion of the gel/membrane cassette in a buffer tank. In commercially available units, the buffer can be chilled and circulated during transfer to eliminate heating artifacts. In semidry transfer *(66, 74)*, the gel/membrane cassette is placed between flat plates that serve as heat sinks.

Reagents

The PVDF membrane of choice should be purchased from the appropriate supplier. 3-[Cyclohexylamino-1-propanesulfonic acid] (CAPS) Ponceau S, Coomassie blue R-250, and Amido black can be purchased from Sigma. Methanol and acetic acid are HPLC grade.

NOTE Wear gloves when handling the membrane to prevent contamination with finger proteins.

Tank Blotting Procedure

1. Cut the membrane to the same size as the gel.

2. Wet the PVDF membrane with a brief immersion in 100% methanol (2–3 sec), rinse with water (2–3 min), and equilibrate in CAPS transfer buffer (10 mM CAPS, 10% methanol, pH 11.0) for at least 15 min.

3. After electrophoresis is completed, rinse the gel in transfer buffer for 5 min. This step is optional. For optimal results with this method, however, equilibrate the gel prior to transfer.

4. Sandwich the gel and PVDF membrane between Whatman 3MM paper and place in the blotting cassette. If desired, a backup sheet of PVDF may be included to capture any protein that passes through the first membrane.

5. Blotting in CAPS transfer buffer is recommended to reduce the level of Tris and glycine contamination from the polyacrylamide gel. The pH is more basic than the p*I* of most proteins and ensures more complete transfer. Deamidation of asparagine is not a concern for two reasons. First, it rarely occurs. Second, deamidation is detected by the appearance of both PTH-Asp and PTH-Asn derivatives in the same cycle. Gels can be transferred equally well in Tris–glycine or other buffers, but the membranes must be thoroughly washed in water to remove these contaminants.

6. The amperage and time of transfer should be optimized for each protein; typically, 20-kDa proteins transfer in 10 min from 0.5-mm-thick gels at 0.5 A (constant amperage). Larger proteins (>100 kDa) may require 45–60 min. Figure 1 shows that small-molecular-weight proteins are recovered on the membrane in higher amounts relative to the high-molecular-weight proteins. Longer transfer times increase the recovery of large proteins at the expense of small proteins (Table I). Soybean trypsin inhibitor is optimally transferred in 15 min under our standard conditions *(84)*. Longer transfer times (30 min) result in decreased recovery on the blot because protein passes through the membrane. A more efficient transfer is obtained using lower amperage, but the transfer times are correspondingly longer *(82)*. Electroblotting is the most critical step in determining the yield of sequenceable protein as 30–80% of the protein loaded in the lane can be recovered on Immobilon-P and up to 100% on Immobilon-P[SQ] and ProBlott.

7. After the transfer is complete, rinse the membrane several times in Milli-Q H_2O (two to three times, 5 min each at room temperature). This step reduces the level of contaminants (Tris and glycine) from electrophoresis and electroblotting to background levels.

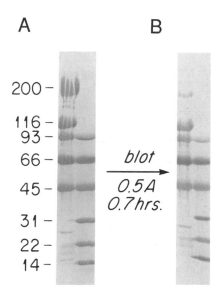

Figure 1 Molecular weight ($M_r \times 10^{-3}$) markers (1 μg each) were loaded in duplicate lanes. (A) Coomassie blue-stained gel and (B) the corresponding blot on a PVDF membrane. Notice the incomplete transfer of the high-molecular-weight proteins. Trace amounts of ^{125}I-labeled molecular weight markers (NEN Dupont) in the sample permit easy calculation of recoveries on the blots and initial yields in the sequence.

Table I

Recoveries of Electroblotted and Electroeluted Proteins

Protein	MW	μg loaded	% on blot	% in eluate
β-Galactosidase	116,250	2.0	85	106
Bovine serum albumin	66,200	2.0	82	87
Carbonic anhydrase	31,000	2.0	89	53
Soybean trypsin inhibitor	21,500	2.0	64	94

Molecular weight markers (DuPont NEN) in Fig. 1 were labeled with ^{125}I and loaded into adjacent lanes for electroblotting (0.5 hr, 0.5 A) and electroelution (400 V-hr). The percentage of protein recovered was calculated from the radioactivity detected by gamma counting.

Semidry Transfer Procedure

1. Cut a piece of membrane and several pieces of Whatman 3MM paper to the same size as the gel.

2. Prepare the transfer buffers. Anode buffer I is composed of 0.3 M Tris, 10% methanol, pH 10.4. Anode buffer II is 25 mM Tris, 10% methanol, pH 10.4. Cathode buffer is 25 mM Tris, 40 mM 6-aminohexanoic acid, 10% methanol, pH 9.4 (40 mM glycine can be substituted for 6-aminohexanoic acid).

3. Wet the PVDF membrane in 100% methanol (2–3 sec), rinse with water (2–3 min), and equilibrate in anode buffer II for at least 15 min.

4. After electrophoresis, equilibrate the gel in cathode buffer for 5 min.

5. Assemble the transfer stack on the anode plate in the following order: Whatman 3MM soaked in anode buffer I, Whatman 3MM soaked in anode buffer II, PVDF membrane, equilibrated gel, and Whatman 3MM soaked in cathode buffer. The number of pieces of Whatman 3MM used for each layer should be as specified by the device manufacturer. A backup membrane can also be included.

6. Transfer the protein from the gel to the membrane. The time and amperage must be optimized for each protein as described previously. Although CAPS buffer can be used in semidry transfers, the overall transfer is less efficient than for Tris buffers.

7. After transfer is complete, rinse the membrane several times in Milli-Q H_2O (two to three times for 5 min at room temperature) to reduce the levels of contaminating Tris and glycine.

Visualization of Proteins

NOTE Use freshly prepared stains and discard after use. Do not reuse stains because contaminants, especially glycine and Tris from the gel, become concentrated in the stain. The staining trays should also be very clean. A convenient tray is the plastic cover from a box of blue or yellow disposable pipet tips.

Coomassie Blue

Visualize the proteins by staining with Coomassie blue R-250 (0.1% in 50% methanol) for 2 min, and destaining with several changes of 50% methanol, 10% acetic acid. Rinse the membrane with several changes of H_2O, airdry, and store at $-20°C$. Blots can be stored for $6-12$ months at $-20°C$ without any effect on the initial or repetitive yields.

Coomassie blue has adequate sensitivity to detect sequence-able amounts of protein. In fact, weakly stained thin bands on a blot represent $50-200$ ng of material. This corresponds to $3.3-13.3$ pmol of a 15-kDa protein or $1-4$ pmol of a 50-kDa protein, amounts not likely to be detected by sequence analysis. **A good test to determine whether there is enough protein for sequence analysis is to photocopy the blot.** If the band is visible on the copy, then generally there is enough material for sequencing. Other dyes like Ponceau S or Amido Black can be used in place of Coomassie blue but are less sensitive. The dyes do not interfere with the sequencing reaction.

Transillumination (72)

Proteins can be visualized without staining by taking advantage of the wetting properties of areas of the PVDF membrane containing blotted protein. After blotting, place the membrane on filter paper and allow it to dry at room temperature. Rewet the blot in 20% methanol and view the blot in front of white light while still wet. Alternatively, place the blot in a glass tray containing 20% methanol and place the tray on a light box. Protein bands will appear translucent against an opaque background. The blot can be photographed if desired. Bands will disappear with drying but can be revisualized by rewetting in 20% methanol. Sensitivity is lower on 0.1-μm pore membranes (e.g., Immobilon-P^{SQ}) than on 0.45-μm pore membranes (e.g., Immobilon-P), 25–50 pmol compared to 5–10 pmol, respectively.

India Ink (98)

India ink staining is recommended for detecting proteins analyzed for phosphoamino acid content because ink does not affect the subsequent acid or base hydrolysis. Place dry membrane briefly in 100% methanol (incubation in methanol for up to 1 min does not cause release of bound proteins), and then incubate for 5 min in 50 mM Tris, 150 mM NaCl (pH 7.4). Incubate wetted membrane for 1.5 hr in 50 mM Tris, 150 mM NaCl (pH 6.5), containing 0.2% Tween-20 and 1 μl/ml Pelican brand India ink.

Relatively short staining times and a slightly acidic pH are recommended to minimize protein loss from the PVDF membrane. Stained membranes are rinsed in pH 6.5 buffer (2 min) without India ink to remove excess carbon particles.

Ponceau S

Dilute the 2% Ponceau S stock solution to 0.2% in 1% acetic acid immediately prior to use. After blotting, rinse the membrane in Milli-Q water. Stain the membrane with 0.2% Ponceau S for 1 min, and wash off excess stain in Milli-Q water (do not exceed 5 min).

Fluorescent Staining (75)

Coull and Pappin modified the transillumination protocol of Reig and Klein *(72)* by adding a water-soluble fluorescent dye to the methanol/water solution. After washing the membrane in deionized water to remove buffer salts, wetting occurs only where protein has been blotted. Binding of the fluorescent dye to the protein bands leaves a permanent pattern that is visible under UV light. Fluorescent staining using this protocol avoids the need to destain while permitting visualization of 10–20 ng of adsorbed protein.

Directly following electroblotting by any of the established methods (e.g., *84, 87*), the PVDF membrane should be washed with deionized water to remove buffer salts (at least two changes of 100 ml, mild agitation of 15–20 min). The membrane is then blotted dry with Whatman 3MM paper and thoroughly dried *in vacuo* for at least 20 min. This drying step is critical for optimum results. The dry membrane is then immersed in a solution of 30% aqueous methanol containing 0.2% (v/v) acetic acid and 0.005% (w/v) (50 mg/ml) of a water-soluble fluorescent dye such as fluorescein or sulforhodamine B (Kodak). After approximately 60 sec of mild agitation, the membrane is washed for a few seconds with deionized water to remove excess dye. The stained protein bands are then visualized by UV light. (With the fluorescein stain, the protein bands are visible only when wet. Sulforhodamine B gives a permanent red stain that remains strongly fluorescent when dry.)

Silver and Gold Staining (69)

Gold and silver staining of proteins blotted to PVDF membranes can detect less than a picomole of sample. Although current sequencing techniques are not sufficiently sensitive to generate useful sequence information from such small amounts of protein, it is possible that direct sequencing of other bands from the same blot might be useful if the mechanism of staining does not interfere chemically with primary amine groups and if colloidal metal does not interfere with the sequencing reaction or detection.

Various amounts of α-lactalbumin (100, 50, 25, and 12.5 pmol) were resolved on a 10–20% polyacrylamide gel and transferred to Immobilon-P using a semidry electroblotter. Following staining with colloidal gold or silver and destaining, the protein bands were excised and subjected to gas-phase sequence analysis *(82)*. No useful data were obtained at any of the protein loadings using either of the metal stains. Therefore, it does not seem possible to obtain sequence information from protein that has been visualized using colloidal metals.

Yields

For Immobilon-P, the amount of protein detectable by sequencing is typically 10–60% of the amount loaded onto the gel. This represents 30–80% recovery of the sample on the blot and 50–80% initial yield from the sequencing instrument. Yields are generally higher on 0.1-μm pore membranes due to enhanced blotting efficiency. The number of amino acid residues identified is dependent on the sequence and the amount of sample, but 10–20 cycles can be obtained from 10–50 pmol of sample.

Poor initial yields reported for this technique usually result from suboptimal transfer. Underblotted gels retain some stainable protein. On Immobilon-P, overblotted gels are devoid of stainable protein but protein can be detected on the backup sheet of PVDF as well as on the back of the primary PVDF sheet. To ensure maximum binding to the membrane either the transfer times must be adjusted or a 0.1-μm membrane should be used. If the protein is insoluble in the transfer buffer, SDS will enhance protein elution from the gel but will interfere with binding to the membrane.

Partial blockage of the N terminus during sample preparation can also decrease the initial yield by as much as 90% *(87)*. N-terminal blockage during electrophoresis and electroblotting can be measured by including a known sequenceable protein such as bovine serum albumin or β-lactoglobulin in an adjacent lane on the gel. In most cases blockage is attributable to impurities in the gel, which can be eliminated by running the gel prior to loading the sample and by including antioxidants such as thioglycolate with the sample.

REMOVAL OF N-TERMINAL BLOCKING GROUPS FROM ELECTROBLOTTED PROTEINS OR PEPTIDES *(150)*

The amino termini of an estimated 80% of proteins in Ehrlich ascites cells are acetylated which prevents direct sequence analysis using the Edman protocols *(146, 147)*. Recent studies have characterized the use of enzymes to remove modified residues from the N terminus of blocked proteins and peptides. If a protein, present in sequenceable amounts, does not yield sequence, it can be subjected to a series of chemical and enzymatic treatments which may remove the blocking group from the N terminus. The protocol for treating blocked proteins is shown in Fig. 2. Normally one discovers a blocked N terminus after the protein is electroblotted to PVDF membranes and subjected to protein sequencing. Do not throw away the membrane! Mild acid treatment can remove the formyl group from f-Met and cause an $N-O$ rearrangement of the acetyl group on serine and threonine residues. Fortunately, these are the most commonly modified amino acids at the N terminus and thus simple chemical treatment is the recommended step to remove an unknown blocking group from a blotted protein. If acid treatment fails to

Nancy LeGendre *et al.*

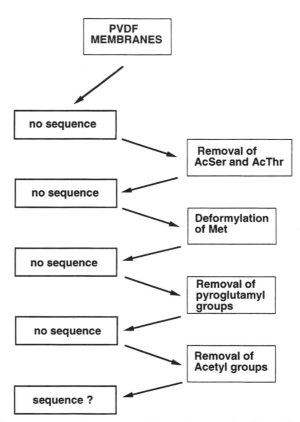

Figure 2 An outline of the sequential steps in removing N-terminal modifications (H. Hirano, personal communication).

generate a sequence, then the blotted protein is incubated with pyroglutamate aminopeptidase, an enzyme that removes pyroglutamyl groups from N terminus of intact proteins. Finally, if these steps fail to generate sequence, then the protein is eluted from the membrane and then subjected to an elaborate series of enzymatic and chemical treatments to remove other acetylated amino acids from the N terminus. Because the enzyme, acylamino acid-releasing enzyme (AARE), cannot act on blotted proteins or peptides larger than 20 residues, the protein must be

86

cluted from the membrane and cleaved into small peptides. The N termini of the peptides are chemically blocked by treatment with PITC and oxidation with performic acid. AARE will remove the acetylated amino acid converting the N-terminal peptide into an *n*-1 peptide whose N terminus can now be sequenced.

Removal of Blocking Groups from *N*-Acetylserine and *N*-Acetylthreonine Residues

Mild acid treatment causes a rearrangement of the acetyl group from the α-amino group to the hydroxyl group of serine and threonine residues *(149,150)*. The time and temperature are minimized to reduce cleavage at internal Ser, Thr, and Asp residues. PVDF membranes are resistant to acids such as HCl and TFA under conditions used in protein hydrolysis for amino acid analysis.

Materials

Obtain TFA (anhydrous trifluoroacetic acid) from Applied Biosystems.

Procedure

1. Remove the band, after several cycles of protein sequence analysis, from the sequencer and place in a 1.5-ml polypropylene microcentrifuge tube.

2. Add anhydrous TFA to the tube and incubate the sealed tube for 1 hr at 40°C.

3. Remove the cap and allow the TFA to evaporate to dryness.

4. Place the band back into the sequencer and subject it to automated analysis.

Removal of Formyl Groups from N-Terminal Met Residues

Materials

Obtain HCl (6 *N* constant boiling temperature) from Pierce Chemical Co.

Procedure

1. Remove the band, after several cycles of protein sequence analysis, from the sequencer and place in a 1.5-ml polypropylene microcentrifuge tube.

2. Add 30 μl of 0.6 *N* HCl to the tube and incubate the sealed tube for 24 hr at 25°C.

3. Remove the cap and remove the HCl with a pipet. Dry the band by vacuum.

4. Place the band back into the sequencer and subject it to sequencing.

NOTE Successful removal of N-terminal formyl or acetyl groups by mild acid hydrolysis critically depends on time, temperature, and concentration of acid. Harsh conditions not only remove the N-terminal blocking group but also cause spurious cleavage in the peptide backbone, particularly at other acid-labile bonds such as Asp-Pro linkages. The background levels of amino acids seen in each cycle are increased under these conditions. In contrast, mild conditions may lead to incomplete removal of the N-terminal blocking groups and very little N-terminal sequence will be detected. The conditions noted above represent a good starting point.

Removal of Pyroglutamate from N-Terminal Residues

Materials

Obtain pyroglutamate aminopeptidase from Boehringer, Pierce, or Takara.

Procedure

Adapted from the supplier's protocol.

1. Remove the band, after several cycles of protein sequence analysis, from the sequencer and place in a 1.5-ml polypropylene microcentrifuge tube. Block the band with 0.5% PVP-40 following the directions described in Chapter 4 for proteolytic digests of proteins blotted on nitrocellulose membranes.

2. Incubate the band with a minimum volume (100 µl) of enzyme (5 µg in 50 mM Na phosphate, pH 7.0, 10 mM DTT) for 5–10 hr at 37°C.

3. Rinse the band with H_2O, dry and place it back into the sequencer.

Removal of Blocking Groups from *N*-Acetylamino Acid Residues *(149, 150)*

Materials

Obtain AARE from Pierce Chemicals or Takara. Prepare performic acid by mixing 9 parts formic acid with 1 part hydrogen peroxide for 1 hr at room temperature. Obtain pyridine and PITC from Pierce Chemicals.

Procedure

1. Remove the band, after several cycles of protein sequence analysis, from the sequencer and place in a 1.5-ml polypropylene microcentrifuge tube.

2. If the membrane has not been blocked with 0.5% PVP-40, follow the protocol in Chapter 4 for *in situ* enzymatic digest on nitrocellulose membranes.

3. After washing the band extensively with H_2O, digest the blotted protein for 24 hr at 37°C with 5–10 µg of trypsin in 100 µl of 0.1 *M* ammonium bicarbonate, pH 8.0, containing 10% acetonitrile.

4. Place the solution into a clean microcentrifuge tube. Wash the band with 100 µl of H_2O and combine the wash solution with the digest eluate. The solution of digested peptides is then dried.

5. React the N termini of the peptides with 100 µl of 50% pyridine and 10 µl of PITC. Purge the mixture with N_2 gas for 20 sec and incubate it for 1 hr at 60°C.

6. Add an aliquot of benzene/ethyacetate (1:1, vol/vol). Vortex the solution and centrifuge for 1 min at 3000 *g*.

7. Remove the supernatant containing the excess reagents and reaction by-products. Extract the lower phase three more times as in step 6. Then dry the lower phase by vacuum.

8. Block the N termini of the peptides by adding 100 µl of performic acid at 0°C for 1 hr.

9. Dry, reconstitute with H_2O, and redry the sample.

10. Reconstitute the peptides with 100 µl of 0.2 *M* Na phosphate, pH 7.2, containing 1 m*M* DTT. Dissolve 50 mU of AARE, add to the buffer, and incubate the mixture for 12 hr at 37°C.

11. Spot the mixture onto modified glass fiber supports and subject it to sequence analysis.

ELECTROELUTION FROM SDS GELS

A second approach to obtaining an internal N-terminal sequence is to digest a protein that is electroeluted from a one- or two-dimensional gel *(83, 102, 103)*. One advantage of electroelution over electroblotting is that the protein cannot be overeluted; however, subsequent concentration and buffer exchange can decrease the amount of protein recovered. The eluted protein can be digested with proteases followed by separation of the peptides by reverse-phase HPLC (Fig. 3)

Materials

Electroelution buffer (Laemmli electrode buffer diluted with an equal volume of water); staining solution (0.1% Coomassie blue

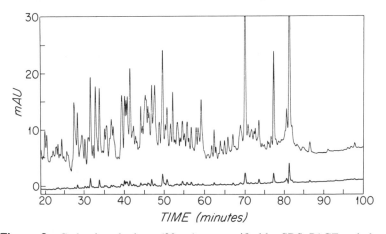

Figure 3   Carbonic anhydrase (20 µg) was purified by SDS–PAGE and electroeluted into a Centricon 30 centrifuge filter. After TCA precipitation and acetone wash, 8 µg was recovered and digested with trypsin. Five micrograms off soluble peptide was injected onto a HPLC narrowbore column (Vydac C_{18}) and eluted following the gradient described in Chapter 2. Upper trace, A_{214}; lower trace, A_{238}.

R-250 in 10% methanol, 0.5% acetic acid); destaining solution (10% methanol); 0.5 ml polypropylene centrifuge tubes (Eppendorf); Centricon 10, 30, or 100 centrifuge concentrator (Amicon), Amicon Centrilutor Microelectroelutor.

Procedure

1. Stain the gel to visualize the bands (10–60 min). The staining and destaining solutions contain reduced concentrations of acetic acid to reduce precipitation of the protein in the gel. Once precipitated, the protein cannot be eluted easily in the absence of SDS. Destain the gel with several changes of destaining solution. A faint band represents approximately 0.5 μg of protein.

2. While the gel is destaining, the microelutor should be assembled. Place a Centricon centrifuge concentrator in the microelutor. The molecular-weight cut-off of the membrane must be smaller than the polypeptide that will be recovered in that particular device. The elutor holds three to eight devices. Put dilute SDS electrode buffer in the buffer chambers of the microelutor. Remove air bubbles trapped at the bottom of the Centricon filters with a pipet.

3. Wear gloves to prevent contamination with finger proteins. After the gel is destained, cut out the bands and place in a 0.5-ml centrifuge tube. The band can be cut into several small pieces. Punch a hole in the bottom and top of the tube with a 20-gauge needle. The band can be stored at –20°C.

4. Place the tube in the top of the Centricon device. Make sure that air is not trapped under the cap of the centrifuge tube. The outside of the tube should be roughened by slitting the plastic with a razor. This ensures that the centrifuge tube fits snugly in the Centricon and does not float out. It is not necessary to have a leakproof fit.

5. The protein is eluted into the bottom of the Centricon tube at 100 V constant voltage for 400–800 V-hr at room temperature. Higher voltages will generate bubbles in the elutor that inevitably block the holes in the tubes. Since the band is stained, the protein can be seen as a dark spot as it accumulates at the bottom of the Centricon. When the gel is clear (all the protein is eluted from the gel), turn off the power and remove the Centricon filters. It takes a minimum of an hour to elute a 10- to 20-kDa protein; a 100-kDa protein requires 4–8 hr. The elution time depends on the voltage, molecular weight of the protein, and the concentration of the polyacrylamide.

6. Concentrate the protein to 50 µl by centrifugation for 20–30 min and recover it from the concentrator according to the instructions that accompany the filters.

7. Precipitate protein in the retentate with 10% cold TCA. Remove SDS with acetone as described in Chapter 2. This step is usually where most losses occur. Keep the supernatants of the washes in case the protein has become soluble or has not precipitated.

NOTE Recent papers [Refs. *(117–119)*] have shown that protein or peptide solutions containing SDS can be separated by reverse-phase HPLC if a DEAE precolumn is used. The DEAE binds SDS which elutes at high acetonitrile concentrations and does not interfere with the separation of peptide mixtures by reverse-phase chromatography. Thus the TCA and acetone precipitation step to remove SDS may be eliminated; the protein sample can be digested in the gel or eluted and digested in solution in the presence of SDS.

8. For N-terminal sequence analysis, reconstitute the pellet with water. Adsorb the protein on a 3 × 6-mm rectangle of Immobilon-P membrane by repeated spotting and drying.

9. For internal protein sequence, the protein can be reduced, alkylated, and digested (described in Chapter 2).

Proteins are eluted and recovered very efficiently because these steps are carried out in the same device. In samples similar to that shown in Fig. 1, 10–20 μg of ^{125}I-labeled molecular weight standards were loaded on a gel. From the recovered radioactivity, it was calculated that 55–106% of the counts in the bands were recovered (Table I). Additional washing steps with cold TCA and acetone (see Chapter 2) resulted in losses as high as 25%. An HPLC narrowbore chromatograph of a trypsin digest of carbonic anhydrase (150 pmol) is shown in Fig. 3. As a conservative rule of thumb, 5 μg of protein is recovered from a 20-μg band. Smaller amounts of starting material result in lower yields, primarily due to nonspecific adsorption to the walls of tubes.

Amino Acid Analysis of Electroblotted Proteins (99)

Amino acid analysis is important for quantifying the amount of protein or peptide present, as well as for determining the most suitable cleavage agent in subsequent digests *(22–25)*. Although samples are generally separated in polyacrylamide gels that contain high concentrations of glycine, it is possible to obtain a reasonably accurate amino acid composition of a protein band from a blot.

PVDF sample blanks should always be run in parallel with the blotted sample. Baseline amino acid content has been shown to vary from lot to lot of PVDF membrane. A true blank for this method would be an equivalent size of PVDF cut from an area of the blotted membrane not exposed to protein. The use of transillumination is preferable to Coomassie staining for protein visualization as Coomassie blue also contributes to artifactual bands and baseline noise.

Procedure

1. Electroblot sample onto PVDF membrane as described previously. Blotting buffers containing glycine contribute background amounts of this amino acid; however, the blots can be washed extensively in deionized water prior to hydrolysis to reduce the level of glycine contamination. Alternatively, glycine contamination can be reduced by using CAPS buffer.

2. Visualize bands by transillumination in white light or with Coomassie blue.

3. Excise the band(s) and blank using a scalpel (typical band 2×10 mm) and place in etch-labeled, acid-washed glass tubes [autosampler insert or Pico Tag (Waters, Division of Millipore) hydrolysis tubes are recommended].

4. Place open sample tubes in a larger glass vessel. The Waters Pico Tag Work Station facilitates multiple sample evacuation; up to 10 samples can be placed in a single reaction vessel. Pipet 200 µl of 7% (w/v) thioglycolic acid in 6 N HCl into the bottom of the outer reaction vessel, and apply vacuum carefully to the closed system. Flush with N_2 and apply vacuum repeatedly to remove all traces of O_2. Add 70 µl of 7% (w/v) thioglycolic acid in 6 N HCl to each sample tube to prevent membrane pieces from flying out of the vial when vacuum is applied.

5. Place the sealed reaction vessel in a 110°C oven for 24 hr.

6. Stop hydrolysis by cooling the reaction vessel to room temperature. Extract amino acids from the PVDF membrane using three 200-µl aliquots of 0.1 N HCl, 30% methanol. Combine the extracts for each sample, and dry under vacuum centrifugation.

Amino acid analysis from PVDF blots has been demonstrated using several derivatization methodologies. For PTC-amino acid analysis, extracts are derivatized *in situ* prior to HPLC separation. Post-column derivatization with ninhydrin (A. Smith personal communication) and *o*-phthalaldehyde have been successfully used, the latter at the 10-pmol level for several blotted proteins.

Phosphoamino Acid Analysis of Electroblotted Proteins (97, 98)

The relative resistance of PVDF membrane to treatment with both acid and base makes it ideal for the analysis of phosphoamino acids in blotted proteins. Partial acid hydrolysis (gas or liquid hydrolysis) is used to analyze total phosphoamino acid content of a protein. Specific detection of phosphotyrosine-containing proteins is facilitated on PVDF blots by preincubation in 1.0 N KOH, which destroys base-labile phosphoserine. Following acid hydrolysis, [^{32}P]phosphotyrosine is identified by two-dimensional electrophoresis.

Sample Preparation

Total cellular protein is prepared by lysis of ^{32}P-labeled cells in hot SDS sample buffer and then fractionated by SDS–PAGE.

Transfer to the PVDF membrane is carried out using Tris–glycine or CAPS buffer.

Stained or unstained PVDF blots should be washed extensively with water (three times for 2 min in 1 liter Milli-Q water) prior to hydrolysis to remove excess glycine, NaCl, and Tween-20 since these components interfere with subsequent electrophoresis. Membranes are dried on Whatman 3MM filter paper under a heat lamp for 5 min. The radioactivity is measured by Cerenkov counting.

Acid Wash for Tubes

1. Soak tubes for 2 hr in 2 N HCl at room temperature.

2. Wash with excess Milli-Q H_2O.

3. Dry.

Vapor-Phase Acid Hydrolysis (97)

1. Place uncapped, acid-washed microfuge tubes in a reaction vial (Waters).

2. Add 1 ml of constant boiling HCl (Pierce) to the bottom of the reaction vial but not the microfuge tubes.

3. Draw vacuum to less than 1 Torr and seal.

4. Incubate at 110°C for 2 hr.

5. Allow vial to cool to room temperature; open and remove microfuge tubes.

6. Wet membrane in microfuge tube with a small volume of 100% methanol.

7. Add 1 ml H_2O to the tube, vortex, and allow sample to soak for 30 min.

8. Remove membrane from the tube by teasing with a micropipette tip. Transfer the membrane to a fresh microfuge tube.

9. Measure the ^{32}P remaining in the aqueous extract and on the membrane by Cerenkov counting (usually > 95% of ^{32}P is eluted from the membrane).

10. Add 10 μl each of phosphothreonine, phosphotyrosine, and phosphoserine (10 m*M* stocks stored at –70°C) and 5 μl 1 mg/ml Orange G (all obtained from Sigma) to the aqueous extract.

11. Freeze and lyophilize.

12. Dissolve in 20 μl H_2O, vortex, and spin briefly in a microfuge.

13. Spot on cellulose TLC plates.

14. Measure ^{32}P in tube after loading plates.

Base Hydrolysis (98)

To detect phosphotyrosine, phosphoserine is removed by base hydrolysis.

1. Detect protein bands with India ink staining and ^{32}P-labeled proteins by autoradiography. Rewet the membranes with methanol and H_2O. It is essential that the membrane be thoroughly wet at this step.

2. Place wetted band in 1.0 *N* KOH and incubate at 55°C for 2 hr.

3. Neutralize by rinsing once for 1 min in 200 μl 50 m*M* Tris, 150 m*M* NaCl (pH 7.4); once for 5 min in 200 μl of 1.0 *M* Tris (pH 7.0); and twice for 5 min in deionized water. Remove excess liquid from the PVDF membrane with Whatman filter paper. Dry the membrane under a heat lamp for 5 min prior to autoradiography.

NOTE   Alkaline treatment of PVDF membranes at elevated temperature changes the color of the membrane from white to brown. Subsequent neutralization will return membranes to a medium tan color. This effect does not interfere with autoradiography and acid hydrolysis. PVDF is entirely resistant to acid treatment.

Less than 10% of bound protein is lost from PVDF blots during alkaline hydrolysis, and membrane-derived autoradiograms are sharper than those obtained from alkali-treated gels. Finally, the phosphoamino acids that remain after partial base hydrolysis of blotted protein can be identified by subsequent acid hydrolysis *in situ*. A sevenfold enrichment of phosphotyrosine over phosphoserine and phosphothreonine is reported for alkali-treated PVDF blots using this methodology.

DESORPTION OF ELECTROBLOTTED PROTEINS FROM PVDF MEMBRANES

SDS–PAGE resolution of proteins can be used as a purification tool prior to sequencing. Direct elution of purified proteins and peptides from acrylamide gels can be accomplished by electrophoresis [*e.g., (77)*] and by diffusion [*e.g., (114)*], but there are a number of problems with both of these protocols. It is difficult to achieve quantitative elution of large-molecular-weight (e.g., >100 kDa) proteins from standard polyacrylamide. Lower density gels with relatively lower levels of crosslinking may still pose sample recovery problems. It is also difficult to excise the gel precisely so that all of the sample is obtained without including contaminating protein from neighboring bands. Additionally, especially with diffusion of proteins out of macerated gels, the protein sample is often recovered in the presence of unwanted salts and polyacrylamide gel. In this case, a second preparative step, such as ultrafiltration, would be required. An alternative to direct elution from the acrylamide gel is electroblotting the sample to a microporous membrane followed by solvent elution of the desired protein off the membrane [e.g., *(95)*]. Staining and localization of sample on a membrane are rapid, and excision of the band can be accomplished by using a scissors or scalpel. Finally, the protein on the membrane is pure and unwanted salts can be washed off easily by rinsing the membrane in the desired solution. A potential problem with desorption of proteins and peptides from microporous membranes, including PVDF, is that the recovery is frequently not quantitative *(96)*. Lower molecular weight hydrophobic samples are more difficult to elute and are generally recovered with lower efficiency. Since the nature and size of fragments that result from proteolytic or chemical digestion are frequently unknown, the difficulties encountered in desorbing a sample are often unpredictable. Szewczyk and Summers *(95)* described a series of methods to desorb different proteins tested (MW range: 14–220 kDa) using a 2% SDS, 1% Triton in 50 mM Tris–HCl, pH 9.0–9.5, as the eluant. Omission of the 2% SDS in the elution buffer resulted in a 10% decrease in recovery for proteins over 200 kDa.

Procedure (95)

After electrophoresis (20 V, ≤20°C, overnight in 25mM Tris/ 192 mM glycine, pH 8.3) the membrane was stained for 20 min using 0.2% (w/v) Amido black 10B in water and then destained in water. The bands of interest were excised with sharp scissors or a scalpel and then placed in an Eppendorf tube containing 0.2–0.5 ml of eluant/cm^2 of membrane. The samples were centrifuged for 10 min at room temperature to release bound proteins from the membrane and then the supernatants were subsequently centrifuged for 5 min at room temperature to remove any remaining insoluble material.

The ability of different eluants to desorb BSA from Immobilon-P is presented in Table II. It may be necessary to dialyze or diafilter protein samples after elution depending on the buffer requirements of subsequent protocols.

Table II
Detergent Elution of BSA from Immobilon-P[a] *(95)*

Eluant	Protein eluted (%)
1% Triton X-100	55
2% SDS	20
20% Acetonitrile	20
1% Triton X-100/2% SDS	100
1% Triton X-100/20% Acetonitrile	100
2% SDS/20% Acetonitrile	100

[a]Base buffer solution for all eluants was 50 mM Tris-HCl, pH 9.0

4 Internal Amino Acid Sequence Analysis of Proteins after *in Situ* Protease Digestion on Nitrocellulose

Ruedi Aebersold

*The Biomedical Research Center
and Department of Biochemistry
University of British Columbia
Vancouver, British Columbia*

Introduction
Gel Electrophoresis
Electroblotting
Protein Detection
In Situ Enzymatic Cleavage of Electroblotted
    Proteins
Reverse-Phase HPLC of the Cleavage Fragments
Peptide Sequence Analysis
Efficiency of the Procedure
Reproducibility of Peptide Maps
Detection of Problems and Troubleshooting

A Practical Guide to Protein and Peptide Purification for Microsequencing, Second Edition
Copyright © 1993 by Academic Press, Inc. All rights of reproduction in any form reserved.

NOTES

INTRODUCTION

The scope of the method is the generation and isolation of peptide cleavage fragments for internal amino acid sequence analysis after separation of low microgram amounts of the intact
polypeptide chain by one- or two-dimensional polyacrylamide
gel electrophoresis. The method most generally recommended
(see flow chart Fig. 4 in Chapter 1) consists of the electrophoretic transfer (electroblotting) of all separated proteins from
the gel onto a nitrocellulose membrane simultaneously, detection of the proteins followed by *in situ* enzymatic cleavage of
individual proteins on the nitrocellulose matrix, and purification
of the released peptide fragments to homogeneity by reverse-
phase HPLC (Fig. 1). These peptide fragments are suitable for

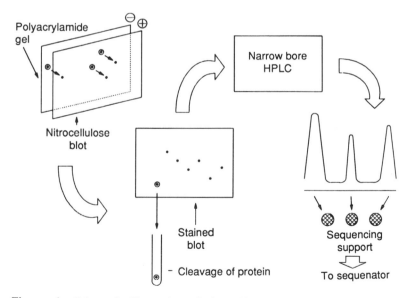

Figure 1   Schematic illustration of nitrocellulose *in situ* protein cleavage
method for the generation of peptide fragments for internal sequence analysis.

use in the gas-phase microsequenator without further manipulation *(100)* or after covalent attachment to suitable supports in advanced solid-phase sequenators *(14, 15)*. Based on the fact that a method is best performed if the principle is well understood, this chapter provides a step-by-step guide through the procedure as actually performed, adds technical details and the rationale behind the manipulations, and suggests measures to troubleshoot the method if problems are encountered.

Materials

For gel electrophoresis, standard commercial sources of chemicals were used and the reagents and solvents were of the highest purity grade commonly available.

Nitrocellulose (0.45 µm pore size) was obtained from Schleicher & Schuell. The protein stains employed were Ponceau S (Sigma) or Amido Black 10B (Sigma). Polyvinylpyrrolidone, average $M_r = 40,000$ (PVP-40) and Polybrene were purchased from Sigma. L-1-Tosyl-amido-2-phenylethyl chloromethyl ketone-treated trypsin, *Staphylococcus aureus* V8 endoproteinase Glu-C (Glu-C) and *Pseudomonas fragii* endoproteinase Asp-N (Asp-N) were all obtained sequencing grade from Boehringer Mannheim. *Achromobacter lyticus* endoproteinase Lys-C (Lys-C) was from Wako Chemicals (Osaka, Japan) and porcine trypsin and chymotrypsin were from Sigma.

GEL ELECTROPHORESIS

Proteins were separated in sodium dodecylsulfate polyacrylamide gels (SDS–PAGE) with the dimensions $70 \times 70 \times 0.5$ mm according to Laemmli or by two-dimensional–PAGE *(62, 64)*.

High purity gel chemicals were used (Bio Rad, Richmond California), and gels were polymerized extensively before use. Even though artificial N-terminal blocking is not a major problem with this procedure, extensive polymerization was included as a precaution to minimize chemical modifications of amino acid side chains. To achieve optimal results, the sample should be applied to the gel as concentrated as possible. Ideally, sample loads approaching the binding capacity of nitrocellulose (5 – 7 μg per 2 × 10 mm band) should be used (see below).

NOTE   Gels with larger dimensions are equally compatible with the procedure. For proteins containing disulfide bonds (essentially only extracellular proteins), it is advised to reduce disulfide bonds and to chemically stabilize the resulting sulfhydryl groups by covalent modification with 4-vinylpyridine prior to gel electrophoresis. A suitable protocol is described in Chapter 2. This reaction is fast, simple and quantitative and yields PTH derivatives that are chemically stable and easily detectable during sequence analysis. Peptide recovery from reduced and alkylated cysteine-containing proteins is generally higher than recovery from nontreated aliquots of the same protein. Pre-electrophoresis treatment generally shows better results than alkylation on the membrane after electrotransfer or in solution after recovery of the cleavage fragments from the matrix (104).

ELECTROBLOTTING

Proteins were electroblotted in a Bio-Rad transblot system onto nitrocellulose for 2 hr from 0.5-mm-thick gels. This procedure is equally compatible with the use of Tris/glycine transfer buffer containing methanol (65) as with borate or CAPS-based blotting buffers. For proteins that were difficult to transfer, up to 0.005% SDS was added to the transfer buffer. Semidry electroblotting systems are compatible with this method. It is essential that conditions given by the manufacturers are followed exactly.

NOTE During electrophoresis in polyacrylamide gels, proteins typically move at speeds of centimeters per hour. Before a protein is absorbed on the support during an electroblotting experiment, the maximal migration distance is the thickness of the gel (typically 0.5–1.5 mm). Comparable field strengths are used for electrophoresis and electroblotting. Therefore, a transfer time of 2 hr is sufficient. Prolonged electroblotting will not generally increase the transfer yields which are not dependent on the size of the protein, provided gels with reasonable pore sizes are used.

Occasionally, poor transfer yields are observed, primarily due to protein precipitation in the gel. During electrophoresis, protein precipitation is prevented by SDS which, through its strongly acidic sulfate groups, also provides the main driving force for protein movement through the gel matrix in an electric field. Unlike the electrophoresis buffer, electrotransfer buffers cannot be supplemented with significant amounts of SDS, because even low concentrations of this detergent prevent protein absorption onto nitrocellulose. In addition, the protein binding capacity of nitrocellulose is increased by the addition of methanol to the transfer buffer, presumably by stripping the SDS coat from proteins. Adding methanol to the transfer buffer aggravates problems with protein precipitation and has a negative effect on the transfer yields of proteins with low solubility. Once proteins are precipitated in the gel they cannot be efficiently transferred, even after prolonged electroblotting.

Although the procedure described in this chapter has been successfully used with PVDF membranes, we prefer the use of nitrocellulose membranes. It appears that the hydrophobic surface of PVDF membranes limits the recovery of peptide fragments. Generally, higher peptide yields are obtained with nitrocellulose. In particular, attempts to perform the *in situ* fragmentation procedure on PVDF membranes after N-terminal sequence analysis of the protein have been unsuccessful.

PROTEIN DETECTION

Transferred proteins were stained with Amido Black *(63)* or reversibly stained with Ponceau S using a modification of a described method *(94)*. For Amido black staining, the nitrocellulose membranes were immersed in a solution of 0.1% (w/v) of Amido black 10B in H_2O:acetic acid:methanol (45:10:45, v/v/v) for 1 to 3 min and rapidly destained with several washes of H_2O:acetic acid:methanol (45:10:45, v/v/v). The destained blots were then thoroughly rinsed in deionized water to remove any excess acetic acid prior to storage at $-20°C$ or further processing. For Ponceau S staining, nitrocellulose filters were immersed for 1 min in a solution of 0.1% Ponceau S dye in 1% aqueous acetic acid. Excess stain was removed from the blot by gentle agitation for 1 – 2 min in 1% aqueous acetic acid. Protein-containing regions detected by either stain were cut out and transferred to Eppendorf tubes (1.5 ml) where Ponceau S-detected protein bands were destained by washing the filter with 200 mM NaOH for 1 – 2 min. Finally, the filter was washed with distilled water and stored wet at $-20°C$, without extensive drying.

NOTE   Both staining procedures did not interfere with proteolytic cleavage, the release of peptide cleavage fragments, or
HPLC analysis of peptides. Both dyes, if released into the supernatant during cleavage, eluted in the void volume of C_4 and C_{18}
reverse-phase columns. The interaction between Amido black
10B and protein was permanent under the mild conditions used
in this procedure. The interaction between Ponceau S and protein is pH dependent and was broken under basic conditions.

The interaction between nitrocellulose and absorbed protein
is also pH dependent and maximal at acidic pH. It is therefore
advisable to perform steps in which protein retention is desired
at acidic pH and steps in which peptide desorption is desired at
basic pH. In particular, the time of removal of Ponceau S from
the protein should be minimized. Extensive drying of the membrane by heat and/or vacuum induced an interaction between
the protein and the support which could not be reversed under
the conditions described.

Protein stains generally show higher staining intensities for
samples electroblotted onto membranes than for equal amounts
of protein in polyacrylamide gels. Protein bands containing less
than 100 ng are detectable on nitrocellulose with Amido black,
whereas the detection limit for Ponceau S is an estimated 200 ng.
This increased staining intensity frequently leads to an overesti-
mation of the amount of protein present in electroblotted samples.

IN SITU ENZYMATIC CLEAVAGE OF ELECTROBLOTTED PROTEINS

Up to five bands (SDS/PAGE) or up to 40 spots (two-dimensional gels) of destained nitrocellulose pieces containing the
same protein (Fig.2) were pooled in a single Eppendorf tube and
incubated for 30 min at 37°C in 1.2 ml of 0.5% (w/v) PVP-40

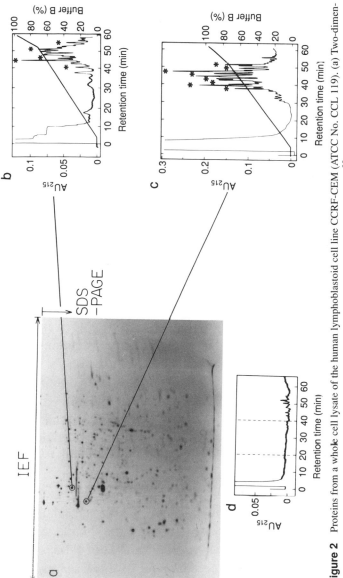

Figure 2 Proteins from a whole cell lysate of the human lymphoblastoid cell line CCRF-CEM (ATCC No. CCL 119). (a) Two-dimensional [isoelectric focusing (IEF)/SDS–PAGE] separation of proteins metabolically labeled with [$^{-35}$s]methionine. Proteins used as examples are circled. (b) HPLC map of peptides released after *in situ* tryptic digestion of the protein ARP after electroblotting onto nitrocellulose. (c) HPLC map of peptides released after *in situ* tryptic digestion of the β-subunit of mitochondrial F$_1$-ATPase after electroblotting onto nitrocellulose. (d) Enzyme blank. Asterisks indicate peptides for which sequence analysis was carried out. AU, absorbance unit.

dissolved in 100 mM acetic acid to prevent adsorption of the protease to the nitrocellulose during digestion. Excess PVP-40 was removed by extensive washing with water (at least five rinses). Because of the strong UV absorbance of PVP-40, complete removal before HPLC analysis is essential. Nitrocellulose strips were then cut into small pieces of approximately 1 × 1 mm and put back into the same tube, rinsed once with 0.5 ml of digestion buffer, and added with the minimal volume of digestion buffer required to submerge the sample pieces.

Proteins were digested on the nitrocellulose by one of the following proteolytic enzymes: Trypsin (bovine or porcine), chymotrypsin, Lys-C, Asp-N, or Glu-C. Cleavage with the proteinases trypsin, chymotrypsin, Lys-C, and Asp-N was performed in 100 mM NaHCO$_3$, pH 8.2/acetonitrile 95:5 (v/v), at 37°C overnight. Cleavage with the proteinase Glu-C was done in 100 mM sodium phosphate, pH 7.8/acetonitrile 95:5 (v/v), at room temperature overnight. The enzyme-to-substrate ratio was kept at a minimum of 1:10–1:20 (w/w). This ratio depends on the amount of protein sample and increases with a decreasing sample load to a maximum of 2:1 for submicrogram amounts of substrate. After digestion, the whole reaction mixture was frozen at −20°C or immediately loaded onto the HPLC column after acidification with 10 µl TFA/H$_2$O 10:90 (v/v).

NOTE Prevention of enzyme adsorption to the nitrocellulose matrix is a key step in this procedure. The "blocking" with PVP-40 is quite effective, but not absolute (Table I). To minimize the effects of enzyme adsorption, the ratio of enzyme to nitrocellulose should be as high as possible. This was best achieved by estimating the actual protein concentration of the sample, e.g., by the use of an analytical, silver-stained minigel of Phast gel. Then 5–7 µg per band of the protein under investigation were loaded onto the gel to be electroblotted onto nitrocellulose. This amount of protein per band (typical size 2 × 10 mm) approached

Table I
Blockage of Nonspecific Protein Binding
to Nitrocellulose

Inhibitor	Inhibition of Binding (%)
None	0
0.5% Ficoll (100 mM)	87
0.1% Tween-20 (100 mM)	71
0.5% PVP-40 (100 mM)	94
0.5% PEG (100 mM)	60
1 mg/ml salmon sperm DNA	−31

Procedure: A 1-cm^2 piece of nitrocellulose was incubated for 30 min at 37°C with the indicated reagents dissolved in 100 mM acetic acid or 1 mg/ml salmon sperm DNA. Reagents were removed by extensive extraction with water. Nitrocellulose pieces were then incubated with 20 µg of ^{125}I-radiolabeled bovine serum albumin at 37°C overnight. Unbound protein was removed by extraction and the amount of bound protein was determined by gamma-radiation counting. The radioactivity was correlated with the values obtained with untreated nitrocellulose (100%). For concentrations in excess of 0.5%, the PVP-40 effect was not concentration dependent.

the maximal protein binding capacity of nitrocellulose. For smaller amounts of protein, the enzyme-to-substrate ratio was gradually increased from 1:10 to 2:1 for submicrogram amounts of substrate.

Particular care should be taken to remove all PVP-40. The inside of the cap of the tube is a likely source of contamination which is easily overlooked. Nonextracted PVP-40 eluted from RP-HPLC columns in the range of 30–40% organic solvent, depending on the buffer system and column used, and was therefore able to obscure eluting peptides. If a multiple wavelength or photodiode array detector is available, PVP-40 is easily identified by its strong absorbance in the range of 215–220 nm and a lack of absorbance at 260–289 nm.

The quality of the proteolytic enzymes is critical for the success of this method. As for digests performed in solution, it is essential to test the enzyme activity prior to setting up important fragmentation experiments. The activity of proteolytic enzymes is somewhat lower for substrates on nitrocellulose compared to substrates in solution.

The initial description of the method evaluated the use of the proteolytic enzymes trypsin and Glu-C (*100*). More recently it has been demonstrated that additional enzymes, including the endoproteinases Lys-C, Asp-N, chymotrypsin, and subtilisin, are equally suited for cleavage of proteins on nitrocellulose membranes (*104*). It is also possible to use different proteases sequentially on the same protein sample to recover peptides from regions of the protein that were not cleaved with the first enzyme used. If bicarbonate is used as the digestion buffer, care should be taken to acidify the sample until no more gas bubbles escape. Some HPLC systems are very sensitive to gas bubbles in the system. Acetonitrile was included in the digestion buffer to enhance the release of peptide cleavage fragments from the nitrocellulose support. None of the endoproteinase activities tested were inhibited by concentrations of acetonitrile of up to 40% (v/v), at which concentration nitrocellulose started to decompose. 5–10% (v/v) acetonitrile was generally included in the digestion buffer since higher concentrations did not further increase peptide release.

REVERSE-PHASE HPLC OF THE CLEAVAGE FRAGMENTS

Peptide fragments derived by enzymatic cleavage were separated on narrowbore (2.1 × 150 mm) Vydac C_4 or C_{18} RP-HPLC columns on a Waters peptide analyzer. This system was equipped with a Rheodyne Model 7125 sample injector with a 500-µl injector loop. The following buffer system was used. Buffer A:

0.1% TFA (sequential grade, Pierce) in water (double glass distilled). Buffer B: 0.08–0.095% TFA in acetonitrile/H_2O, 70:30; (v/v). The absorbances of buffers A and B were matched at 215 nm by titrating the TFA concentration in buffer B. Both buffers were degassed with a stream of helium. Peptide separations were generally carried out at a column temperature of 50°C at a flow rate of 100 µl/min. Sample application was followed by a 5 min isocratic wash of the column with 100% solvent A at a flow rate of 200 µl/min. This initial phase was followed by a linear gradient from 0 to 70% acetonitrile. Peptides were detected at a leading wavelength (216 nm) and two secondary wavelengths (260 and 280 nm) using a photodiode array detector. Because low picomole quantities of peptides (nanogram amounts) are generated, high sensitivity UV detectors are required.

Fresh or thawed peptide-containing solution was acidified with 10 µl of 10% TFA, mixed quickly by vortexing, and centrifuged for 1 min in a microfuge at high speed. The supernatant was removed and the nitrocellulose pieces were washed once with 20–50 µl of 0.1% TFA in H_2O. The combined supernatants were injected onto the HPLC column.

Peptides were collected manually into plastic microfuge tubes according to UV absorbance at 215 nm. In cases in which the shape of the peptide peak suggested co-elution of two or more peptide species (Fig. 3a), the collected sample is rechromatographed using a separation system with different selectivity. The volume of the collected fractions was reduced in a SpeedVac vacuum concentrator to 20 µl. The sample was then mixed with 20 µl of 0.1% TFA in water and injected into the pre-equilibrated RP-HPLC system equipped with a column of different chromatographic selectivity. Individual components were separated by developing the columns with a linear gradient of acetonitrile (Figs. 3b and 3c). Solvents used for the second chromatography dimension included 0.1% phosphate, 150 mM NaCl in H_2O, 0.1% ammonium acetate, and 0.1% hydrochloric acid.

Ruedi Aebersold

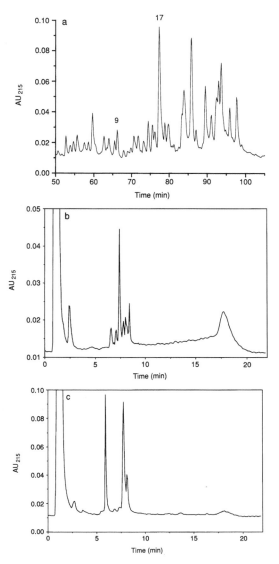

Figure 3 Separation of co-eluting peptides by rechromatography under conditions with different selectivity. Selected peptides collected from primary RP-
HPLC separation in the TFA/H$_2$O/CH$_3$CN solvent system (a) were separated in
the Michrom UMA microbore HPLC system using a 1 × 5-mm PRLPS column
and a solvent system containing 0.1% HCl. Rechromatography of peak No. 9 (b)
and No. 17 (c) are shown. AU, absorbance units.

116

NOTE The volume of the peptide-containing solution eluted from the nitrocellulose membrane usually was relatively large (25–100 µl). To efficiently wash buffer salts, excess dye, and other contaminants through the column, the HPLC system was best run isocratically for one system volume, before the gradient was started. The isocratic initial phase leads to the loss of very short, hydrophilic peptides, which do not stick to the column. To ensure that all the injected sample was loaded on the column, the volume of the injection loop was chosen to be at least twice the volume of the injected sample.

Peptide-containing fractions are collected manually into Eppendorf tubes based on the UV absorption at 215 nm. Collected fractions are frozen immediately. If peptides are not frozen immediately after the collection, the yield of recovered phenylthiohydantoins drops significantly at each serine residue during sequence analysis.

Rechromatography of peptides collected from the primary RP-HPLC separation was an essential step in minimizing the chance of sequencing mixtures of co-eluting peptides (Fig. 3). Using a standard chromatography system, the peptide loss associated with rechromatography is typically in the order of 30–40%. Using a specialized micro-bore HPLC system (Michrom, UMA, Michrom Bio Resources, Inc., Livermore, CA), near quantitative peptide recovery can be achieved. Sequence analysis of the collected peptides in the gas-phase or pulsed-liquid phase sequencing mode was compatible with all the solvent systems described for rechromatography, although the use of phosphate is not recommended because phosphate tends to accumulate in the sequencer and increasing amounts of phosphate progressively interfere with the Edman degradation. Covalent attachment of collected peptides for solid-phase sequence analysis through carboxyl groups by carbodiimide *(14, 15)* is not compatible with the presence of acetate. Any solvent system containing acetate should therefore be avoided for the isolation of peptides for covalent sequencing.

117

PEPTIDE SEQUENCE ANALYSIS

Amino acid sequence analysis of peptides prepared by the described procedure can be performed on any gas–liquid phase protein sequencer using conventional Edman chemistry (Fig. 4). Peptides prepared by *in situ* cleavage are equally compatible with advanced solid-phase sequencing protocols *(14, 15)* and with sequence analysis by mass spectrometry. This is of partic-

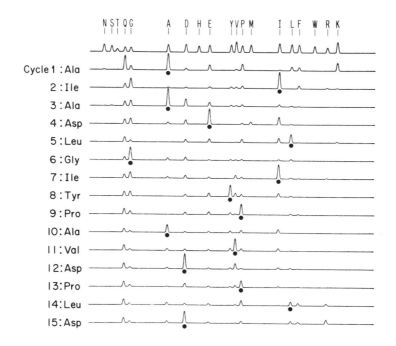

Figure 4   Sequence analysis of tryptic peptides (peak No. 8, Fig. 2c) of F_1 ATPase β-subunit. Peptide peak No. 8 was collected from narrowbore reverse-phase HPLC and applied to the cartridge of a gas-phase protein sequenator. HPLC analysis of PTH residues in cycles 1–15 are shown. *Top*: Mixture of PTH standards, 10 pmol each. PTH derivatives are designated by the single letter code of corresponding amino acids.

ular importance since these two procedures are ideally suited for the characterization of modified amino acid residues.

EFFICIENCY OF THE PROCEDURE

To quantitate the efficiency of the procedure, 10 μg of ^{125}I-radiolabeled α-lactalbumin was applied to 1-cm^2 pieces of nitrocellulose. The protein was stained with Ponceau S and destained, and the nitrocellulose pieces were treated with 0.5% PVP-40 solution. Digestion with trypsin was carried out for 12 hr at 37°C and the resulting recovered peptide fragments were separated by RP-HPLC. The efficiency of each step of the procedure was determined by gamma radiation counting.

Although α-lactalbumin is a relatively small and "nonsticky" protein and represents an unfavorable case in terms of washout losses, the overall recovery of peptides amounted to 65% of the protein initially present on the nitrocellulose (Table II). Equally high or higher yields have been reported with a wide variety of proteins and proteolytic enzymes *(104)*.

Several parameters proved to be critical for the high overall efficiency of the procedure. (i) Proteins are more strongly adsorbed to nitrocellulose at acidic pH than at neutral or basic

Table II
Efficiency of Internal Protein Sequencing Method

Protein on nitrocellulose	100%
Left after staining/destaining	96%
Left after PVP-40 "saturation"	83%
Release after tryptic digest	77%
Adsorbed to tube	14%
Overall recovery of peptides	63%

119

Ruedi Aebersold

pH. If the PVP-40 treatment was performed at acidic pH, very little protein was washed off during saturation. (ii) Hydrophobic interaction between peptide fragments and nitrocellulose can generally be reduced by adding organic solvent. Thus, 5% acetonitrile was included in the digestion buffer to enhance the release of the peptide fragments. (iii) The yield was higher if the protein concentration on the nitrocellulose blot approached the maximal binding capacity. (iv) It is essential that during the whole procedure the nitrocellulose blots never completely dry out or are exposed to heat.

REPRODUCIBILITY OF PEPTIDE MAPS

Enzymatic peptide maps of the same protein digested either *in situ* on nitrocellulose blots or in solution are somewhat different, although most of the major peptides can be found in both preparations (Fig. 5). Parallel processing of multiple electroblotted samples of α-lactalbumin show that the digestion pattern is reproducible.

The difference in the peptide maps after enzymatic fragmentation *in situ* and in solution probably reflects differences in accessibility of enzymatic cleavage sites or differential release of peptides. The overall recovery, estimated from the relative peak areas in HPLC analyses of *in situ* and solution digests, correlates with the overall efficiency of 50–60% as calculated from quantitative experiments using radiolabeled proteins. Late eluting peaks tend to be absent in separations of peptides derived by enzymatic fragmentation on nitrocellulose, suggesting that very long and hydrophobic peptides are retained on the matrix. Separation of peptide mixtures recovered from nitrocellulose do not show any contaminating peaks derived from

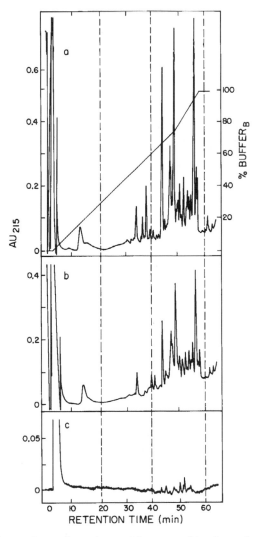

Figure 5 Comparison of tryptic peptide maps of bovine α-lactalbumin digested in solution or on nitrocellulose. Aliquots of 10 μg of bovine α-lactalbumin were either directly applied to a tube for tryptic cleavage in solution or subjected to SDS–PAGE, electroblotting onto nitrocellulose, and *in situ* tryptic cleavage. The resulting peptide cleavage fragments were separated by reverse-phase HPLC. (a) Cleavage in solution, (b) cleavage on nitrocellulose, and (c) trypsin blank.

121

staining or PVP-40 saturation procedures or from the nitrocellulose itself. A protease blank reaction is always included to unambiguously identify substrate-derived peptides.

DETECTION OF PROBLEMS AND TROUBLESHOOTING

Table III gives examples of problems that might be encountered, possible diagnoses, and recommendations for therapy.

Table III

Symptoms	Diagnosis	Therapy
Proteins left in gel after electroblotting	Proteins probably precipitated in the gel	Try adding small amounts (<0.01%) of SDS to transfer buffer Try softer gel
After blotting, proteins not in gel, and not on blot	Protein lost	Check polarity of blotting setup (proteins should move toward + pole) (red)
After blotting, no protein is left in gel, but protein bands only stain faintly on blot; front and backside are stained with comparable intensities	Protein moved through the filter, or protein was never in gel Low binding capacity	Check electrophoresis system Check sample protein concentration in analytical gel Reduce detergent load in blotting buffer (eliminate addition of SDS, use thinner gel, soak gel in transfer buffer for 15 min before blotting) Check whether methanol has been added to transfer buffer Check pH of blotting buffer

(continued)

Symptoms	Diagnosis	Therapy
Staining intensity of protein bands of nitrocellulose is weaker than expected	Not enough protein was loaded Fraction of the protein moved through the membrane	Check protein contents in solution with analytical gel or determine protein concentration (see previous)
Protein bands appear fuzzy on the filter	Inefficient electrophoresis system Air bubbles between gel and nitrocellulose during transfer	Check electrophoresis system Remove air bubbles by carefully moving glass pipet over membrane before blotting sandwich is assembled
After injection of the peptide solution into HPLC, recorder pen goes off-scale and does not come back down	Air bubbles	Apply moderate back pressure to HPLC system (Caution: do not break flow cell) Wash column with 100% solvent B Degas next sample before injection
No peaks appear in HPLC, flat baseline	No solvent gradient or flow	Check HPLC
No peaks appear in HPLC, but recorder shows characteristic baseline rise	Solvent gradient is delivered, no sample is injected	Check injector orientation and setup
Within minuter after injection, big solvent peaks appear, but no peptide peaks are detected	No enzymatic cleavage; the fact that there are no peptide peaks at all (not even from autodigested enzyme) suggests that there was no enzyme present or it was absorbed into the membrane	Check whether enzyme was added Check whether PVP-40 treatment was successful Increase protein load per unit area of the membrane
Peptide peaks are smaller than expected or only peptides from autodigested enzyme are present	Enzyme was present, not absorbed, but not active	Check enzyme activity in solution digests (compare a number of enzyme batches, choose the best) Check buffers used during digestion Make up fresh enzyme solutions and digestion buffer daily

(continued)

123

Ruedi Aebersold

Symptoms	Diagnosis	Therapy
Sharp, offscale peak around 30–40% acetonitrile in HPLC analysis	Most likely PVP-40. (characteristic for PVP-40 is its strong UV absorbance at 210–220 nm and the lack of absorbance at 260–280 nm)	Improve PVP-40 extraction Check whether droplet of PVP-40 solution was left in tube's cap
Even after extensive and careful extraction, an offscale contaminant peak obscures eluting peptides	Most likely degradation product of PVP-40 or contaminant associated with some batches of PVP-40 (this nonremovable contaminant has been reported to N-terminally block peptides[a])	Use new batch of PVP-40 Make up PVP-40 solution daily
Poor peptide separation	Insufficient HPLC system	Check column Check buffer Check HPLC
Peptide loaded into sequenator, but no sequence	As a rule of thumb a peptide peak whith an absorbance at 215 nm of 0.01–0.02 should be easily sequenceable (10–30 pmols PTH signal). If yields are lower, sequencing may be inefficient or peptide is lost	Check efficiency of sequenator with standardized peptide sample Check with standardized peptide sample, whether amino terminus gets blocked under HPLC conditions used Never dry down peptide in tube
Significant drop in sequencing yields after serine residues	Modification of serine amino acid side chain leading to partial blocking	Freeze collected peptide solution immediately after collection from HPLC

[a]R. Frank, DKF2, Heidelberg, Federal Republic of Germany, personal communication.

5 Mass Spectrometric Strategies for the Structural Characterization of Proteins

Hubert A. Scoble, James E. Vath, Wen Yu, and Stephen A. Martin

Genetics Institute, Inc.
Structural Biochemistry Department
Andover, Massachusetts

Introduction
Mass Spectometric Instrumentation for Biological
  Analysis
    Matrix-Assisted Laser Desorption Time-of-Flight
      Mass Spectrometry
    Electrospray Mass Spectrometry
    Fast Atom Bombardment Mass Spectrometry
    Tandem Mass Spectrometry
Application of Mass Spectrometric Approaches
  to the Structural Characterization of
  Recombinant Glycoproteins
Conclusions

NOTES

INTRODUCTION

In recent years, mass spectrometry has found increased application for the primary structural characterization of proteins and glycoproteins. To date, most biological mass spectrometry applications have involved the use of FAB MS for the precise and accurate determination of the molecular mass of peptides resulting from proteolytic digestion or specific chemical degradation of a protein (121, 122). With some prior knowledge regarding the putative structure of a protein, mass spectrometric approaches are unequaled for speed of analysis, ultimate sensitivity and the ability to simultaneously analyze complex mixtures of peptides. For these reasons, mass spectrometry has become indispensable for the structural characterization of recombinant proteins (123, 124). Structural questions concerning post-translational processing or modification can many times be answered by the determination of the molecular mass of peptides from the protein under investigation. Thus, the wide variety of mass spectrometry applications have largely focused on the confirmation or correction of DNA-derived protein sequences, with specific attention being paid to post-translational processing and covalent modification.

From the viewpoint of primary structural analysis, the molecular mass of a peptide is of limited value without additional information. Although it may be possible to calculate amino acid compositions that are consistent with the experimentally determined molecular mass of a peptide or protein, the possible compositions grow exponentially with increasing mass, and even if available, these compositions say little about the linear arrangement of amino acids within the molecule. The development of tandem mass spectrometry has provided the opportunity for mass spectrometry to contribute to the solution of structural questions where prior knowledge concerning the protein structure is not a prerequisite *(125, 126)*. The structural information generated by tandem mass spectrometry is comparable to that obtained by N-terminal sequence analysis, albeit by very different technologies. The potential advantages of MS-based approaches for peptide sequencing are numerous; however, the widespread application of tandem mass spectrometry to peptide sequencing has been limited. This has been due, in part, to several factors, including the complexity, limited availability, and expense of the instrumentation, as well as a general belief that the sensitivity of analysis was not comparable to that of automated N-terminal sequence analysis. Thus, applications of tandem mass spectrometry have focused on structural questions that were difficult or impossible to approach conventionally. Such questions as the structural characterization of N-terminally blocked peptides and the location and identification of sites of *N*- and *O*-linked glycosylation, as well as situations in which the peptide under investigation has been otherwise post-translationally modified or processed are those that have been most commonly addressed mass spectrometrically.

More recently, two additional mass spectrometric techniques, MALD-TOF MS *(127)* and ES MS *(128)*, have found increased application for protein and glycoprotein structural analysis. These techniques are primarily recognized for their ability to generate accurate molecular mass information on intact proteins and glycoproteins. These molecular mass measurements are based on the chemical properties of the protein and are therefore minimally influenced by the protein hydrodynamic or physical properties. Since these are emerging mass spectrometric techniques, their application to biological structural issues is not widespread.

The choice of an appropriate mass spectrometric approach is largely dependent on the structural question under consideration. As briefly discussed previously, matrix-assisted laser desorption and electrospray are used for the determination of the molecular mass of intact proteins, FAB and electrospray for accurate mass assignments of peptides and small proteins, and MS/MS for primary structure determination of peptides or glycoconjugates. Common to all these mass spectrometric techniques is the need to form stable gas-phase ions from the sample of interest. This feature, combined with specific sample issues (e.g., quantity, molecular mass, purity, microheterogeneity) may limit the practical utility of one mass spectrometric technique compared to another. This chapter briefly describes a number of mass spectrometric based techniques that have or will have a significant impact on the current classical approaches for protein and glycoprotein structural characterization. The majority of the discussion in this chapter focuses on the analysis of positively charged molecules, $(M+H)^+$; however, it may be advantageous under certain circumstances to analyze negatively charged ions, $(M-H)^-$. The choice of positive or negative ion detection depends primarily on the analyte and the structural question under investigation. All the techniques discussed in this chapter require the mixing of the sample with a substrate referred to as the matrix. The presence of the matrix is a prerequisite for the formation of ions related to the sample. The basic principle and operational parameters, as well as sample requirements for each of these techniques are detailed below and briefly summarized in Table I.

Hubert A. Scoble *et al.*

Table I

Characteristics of Biopolymer Ionization Methods

Instrument Characteristics	MALD	ES	FAB	MS/MS
Sensitivity (pmol)	0.5–5.0	0.5–1.0	20–1000	100–2000
Mass limit (daltons)	200,000+	50,000+	7000+	2500+
Mass accuracy (%)	0.1	0.01	0.005	0.005
Resolution	200+	5,000+[a]	10,000+[a]	3,000+[a]
Acquisition time (min)	1–2	1–5	2–5	2–5
Sample requirements				
Concentration (pmol/μl)	5	5	100	500
Volume consumed (μl)	2	20[b]	1	1
Matrix/Solvent[c]	Sinapinic acid/H_2O /CH_3CN	2% acetic acid/H_2O /CH_3CN	Glycerol/ H_2O	Glycerol/ H_2O
Purification	No	Yes	Yes	Yes
Interpretation time	Minutes	Minutes+	Minutes	Hours

[a] Double focusing mass spectrometer.
[b] Flow rate = 1-20 μl/min.
[c] See Table II for details.

MASS SPECTROMETRIC INSTRUMENTATION FOR BIOLOGICAL ANALYSIS

Matrix-Assisted Laser Desorption Time-of-Flight Mass Spectrometry

MALD-TOF MS, as first described by Hillenkamp *et al. (127)*, is an extremely accurate and sensitive technique for the determination of molecular masses extending to greater than 200,000 Da. The technique involves the mixture of a solution of a chromaphoric matrix with a biopolymer in a molar ratio of

1000:1–10000:1. Experimentally, this translates to a few pico-moles of protein per analysis. The analyte and matrix mixture is deposited and dried on a sample stage, inserted into the mass spectrometer and irradiated with a pulsed laser at an absorption maximum of the matrix. The interaction of photons with the matrix and sample results in the formation of intact ions related to the molecular mass of the biopolymer. Although matrix-assisted laser desorption ionization may be coupled to a variety of mass spectrometers *(130, 131)*, it is most often coupled to time-of-flight mass spectrometers *(127)*. In this instrument the mass is calculated based on the time required for the biopolymer to travel from the point of ion formation to the point of detection. A schematic illustrating the principles of operation is shown in Fig. 1. This technique has a demonstrated sensitivity of ≈5.0 pmol for the mass range extending to 200,000 Da, a typical mass accuracy of 0.1% and a sample acquisition time of a few minutes.

$$m\,/\,z \,\propto\, (time\text{-}of\text{-}flight)^2$$

Figure 1 Schematic of the basic principles of matrix-assisted laser desorption time-of-flight mass spectrometry. The laser irradiates the sample surface with a fixed wavelength, desorbing three different molecular entities. The lightest of these molecules (c) reaches the detector in the shortest amount of time while the heaviest molecule (a) arrives last. The exact masses of molecules a, b, and c are determined based on their discrete travel times from the ion source to the detector. The mass to charge ratio (m/z) of the desorbed molecule is proportional to the square of the fight time.

A key feature of MALD-TOF MS is its ability to rapidly determine the chemical molecular mass of a biopolymer with more accuracy than that provided by gel electrophoresis. Typically, errors on the order of 10 Da in 10,000 Da or 50 Da in 50,000 Da are attainable *(132)*. Since MALD-TOF MS is a measure of the chemical mass of the molecule, it is unaffected by the physical or hydrodynamic properties of the protein. In comparison, protein migration in SDS–PAGE is dependent on the amount of SDS bound to the protein and its subsequent migration in an electric field. Additionally, protein mobility in the gel system may be very sensitive to protein conformational changes and insensitive to protein or carbohydrate structural changes. In general, protein mobilities are difficult to predict a priori and mobility differences are sometimes difficult to interpret based on the shift in mass. Another feature of MALD, which distinguishes it from other ionization methods discussed in this chapter, is that molecular masses may be detected in the presence of a wide range of buffer components *(133)*. This relative insensitivity to buffers means that less sample handling and fewer purification steps are required prior to mass spectrometric analysis. Furthermore, although a wide range of matrix/laser combinations (Table II) yields a molecular mass for a given biomolecule, it appears that certain matrices may work better than others

Table II

Common Matrices

Technique	Matrix
FAB/LSI	Glycerol, 1-thioglycerol, 2-nitrobenzyl alcohol, dithiothreitol:dithio-erythritol (5:1, w/w), acetic acid
MALD	nicotinic acid (266 nm), sinapinic acid (337 nm), 2,5 dihydroxyben-zoic acid (337 nm), α-cyano 4-hydroxy cinnamic acid (337 nm)
ES	H_2O/organic (CH_3OH, CH_3CN, etc.) w/o 2% acetic acid or NH_4OH

depending on the compound type *(127)*. Based on initial reports, this technique appears to be able to detect the most diverse range of biopolymers, including proteins and glycoproteins *(134)*, carbohydrates *(135)*, and nucleic acids *(136)*; however, the key parameters for successful sample preparation have not yet been fully investigated.

Electrospray Mass Spectrometry

Another technique that provides molecular mass determination for intact proteins is electrospray ionization *(128)*. In contrast to the solid sample surfaces of MALD, ES is a liquid introduction technique in which the sample ions are formed as a result of spraying the liquid into a region of high electric potential at atmospheric pressure. In this process the droplets are desolvated and the resulting gas-phase ions can be mass analyzed by a variety of mass spectrometers. The desolvation and ionization process are illustrated in Fig. 2. A unique aspect of electrospray ionization is the formation of a number of signals in the mass spectrum which result from the multiple charge states of the molecule. All of these multiply charged ions can be correlated to the molecular mass of the sample by determining the observed mass and the charge state of each ion *(128)*.

Sample consumption in this technique is similar to that in MALD, however, sample preparation and introduction are different. Unlike MALD, the sample is introduced as a liquid into the mass spectrometer at a flow rate of $1-5$ µl/min with a sample concentration ranging from 1 to 20 pmol/µl. The total data acquisition time is sample dependent ranging from several seconds to several minutes. As in MALD, the choice of matrix influences the quality and extent of information obtained from the sample. In ES the matrix is the solvent system. The most com-

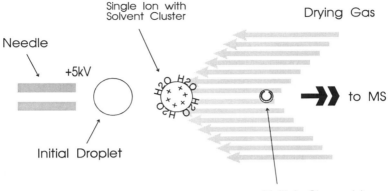

Figure 2   Schematic of the basic principles of electrospray ionization mass spectrometry. The sample dissolved in a water/organic mobile phase sprays from the tip of a needle set at a high potential (+5 kV) into a chamber that is at atmospheric pressure. For simplicity, a single droplet from the fine mist is depicted. The droplet, which contains a fixed number of charges, is gradually desolvated by the countercurrent flow of a warm drying gas resulting in the formation of multiply charged ions. These ions are extracted into the high vacuum region of the mass spectrometer for mass analysis.

mon system for peptides and proteins is a volatile buffer consisting of water/acetonitrile containing a small amount of trifluoroacetic acid and/or acetic acid. These solvents are commonly used in reversed-phase HPLC. As in MALD, the effects of various matrices (i.e., solvent systems) have not been fully explored because this technique is still in the early stages of development and refinement. Since this technique is directly compatible with liquids, it can be coupled with chromatographic systems for on-line analysis *(137)*. Furthermore, this technique is capable of detecting molecular masses ranging from single amino acids to intact proteins and for sequencing peptides *(138)*.

Fast Atom Bombardment Mass Spectrometry

Although the two previously mentioned techniques have a dynamic mass range extending from a few hundred to several hundred thousand daltons, they are not currently as widely available as FAB, also referred to as LSI MS *(139)*. This ionization method, which was introduced in 1980 *(140)*, led to the widespread use of mass spectrometry for biological structural analysis. Prior to the introduction of FAB, most mass spectrometric strategies for the analysis of peptides and proteins involved extensive digestion of the molecule yielding primarily mono- to pentapeptides. The resultant peptides were derivatized and analyzed by gas chromatography–mass spectrometry *(141)*. FAB was the first widely available technique for the analysis of underivatized peptides and small proteins.

The key feature of this technique, as in MALD and ES, is the matrix (Table II). In contrast to MALD and ES, sample preparation has been studied in detail for more than a decade *(142)*. In FAB, the matrix compounds must exist in the liquid phase at very low pressures that are required in the high vacuum region of the mass spectrometer. The most commonly employed liquid matrices for FAB are glycerol and thioglycerol. Sample preparation consists of mixing the peptide with a few microliters of glycerol such that the final protein concentration is in the region of 10^{-4} to 10^{-5} M (10–100 pmol/µl) and the final volume is no more than twice that of the total glycerol volume. In most instances the initial sample volume is much greater than the final 1–2 µl of glycerol. In these cases the glycerol is added to the larger volume and the sample is mixed and concentrated by vacuum centrifugation to remove the excess solvent. The final solution volume consists primarily of glycerol with the dissolved sample. If the sample is isolated from a reversed-phase

Hubert A. Scoble *et al.*

HPLC separation, fractions should be dried to less than half their initial volume prior to adding the matrix. This reduces the probability of a reaction occurring between the matrix and any HPLC eluent components. As in electrospray, the addition of a dilute acid (e.g., 5% acetic acid) to the matrix may enhance the mass spectrometric response for the sample in the positive ion mode.

Once the sample is prepared, 0.5–1 μl of the solution is deposited on the surface of the sample stage and inserted into the mass spectrometer. The sample is bombarded by a beam of high energy atoms or ions, ejecting sample and matrix ions (FAB ion source, Fig. 3) which are mass analyzed and detected with a typical mass accuracy of several tenths of a dalton. The principal species found in a FAB mass spectrum are related to the molecular mass of the peptide with the addition of one proton to give it a net single positive charge, i.e., $(M+H)^+$. As noted previously, in certain situations it may be more advantageous to detect negative ions.

An important advantage of FAB is that a simple mixture of peptides with similar solubility characteristics can be analyzed and detected simultaneously as long as each species has a different molecular mass. This scenario often occurs during the analysis of peptides partially resolved by reversed-phase chromatography. MALD and ES can also analyze mixtures of species which differ in molecular mass. A disadvantage of FAB, which is shared with MALD, is the presence of a large number of ions associated with the matrix below mass 300, making it very difficult to identify sample ions in this low mass range. Furthermore, in contrast to MALD and to a lesser extent ES, FAB is extremely sensitive to buffer salts (e.g., Li, Na, K) and surfactants. The presence of these species often severely reduces or eliminates any ions corresponding to the molecule of interest, therefore, the vast majority of these samples are purified by reverse-phase HPLC with volatile buffer components prior to analysis.

136

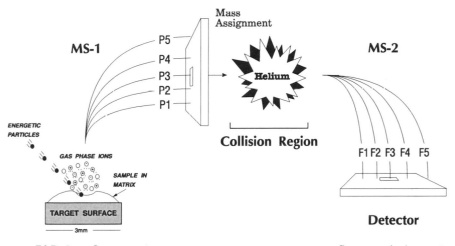

FAB Ion Source

Sequence Assignment

Figure 3 Schematic of the basic principles of FAB ionization, mass spectrometry, and tandem mass spectrometry (MS/MS). The sample, dissolved in a liquid matrix, is desorbed by energetic particles (atoms or ions) in the FAB ion source and the resulting gas-phase ions are separated based on their mass in the first mass spectrometer (MS-1). In this figure, five peptides of different mass (P1–P5) are detected in the sample. In the MS/MS mode, a single mass corresponding to a peptide of interest, in this example P3, is transmitted into the collision region where it collides with the helium gas. The P3 ion fragments into a number of product ions (F1–F5), which are mass analyzed by the second mass spectrometer (MS-2). The primary sequence of the peptide is deduced from these data.

Tandem Mass Spectrometry

MALD, ES, and FAB are all methods to form gas-phase ions from samples that may be in the liquid or solid state in the presence of a variety of matrices. These techniques primarily generate ions that are related to the molecular mass of the sample, and in the standard instrument configurations, only molecular mass information is obtained. These data provide little, if any, information concerning the identity or sequence position of amino

137

acids in the peptide or protein. Primary structural information may be obtained by combining any of the above-mentioned ionization methods with a technique referred to as tandem mass spectrometry *(125, 126)*. In this technique, two stages of mass analysis are linked in series. In the first stage, molecular ions are formed and subsequently separated based on mass. The sample ions at the molecular mass of interest are then transmitted into a region between the two analyzers (collision region) where they are broken apart into a variety of subspecies (fragments) which are subsequently mass analyzed and detected in the second analyzer. The fragmentation of these intact peptide ions into subspecies, i.e., product ions, is most often accomplished by collision with an inert gas such as helium. This process is illustrated in Fig. 3 for a sample containing a mixture of five peptides (P1– P5).

The fragmentation process produces a distribution of product ions ranging from masses corresponding to the core structure of the amino acids, to masses corresponding to series of amino acids from 1 to N, where N is the total number of amino acids in the peptide. Although the collision process is random, sequence-specific fragment ions are routinely detected, enabling the assignment of the complete amino acid sequence for the majority of peptides where the molecular mass is < 2000 Da. These assignments are based on extensive investigations of the fragmentation of peptides in the late 1980s *(126, 143, 144)* from which a number of rules and guidelines have been elucidated. Two unique features of tandem mass spectrometry for determining primary sequence include the ability to (1) sequence peptides in the presence of other peptides *(125)* and (2) unequivocally establish the sequence of N-blocked peptides or to determine location and structure of peptides that are otherwise post-translationally modified *(125)*. The former advantage is achieved whenever the peptides in a mixture have different molecular masses because only one mass is selected for transmission into the collision region. Therefore, all subspecies formed upon collision will be related only to ions of the molecular mass that was selected in the first analyzer. The second advantage is realized because the formation of molecular ions identifies the mass difference associated with the modification, and the subsequent fragmentation into product ions yields ions associated with both the expected amino acids as well as any structural modification enabling the assignment of the complete sequence.

This technique typically requires more material than is necessary to obtain the molecular mass of the molecule (100 pmol to several nanomoles). The data acquisition time is similar to that of the other techniques; however, the complete interpretation of an MS/MS spectrum may take a few hours depending on the complexity. Furthermore, the mass range is more limited in the MS/MS mode compared to molecular mass analysis, with a practical upper mass range limit of 3000 to 3500 Da for obtaining reasonably complete sequence information.

APPLICATION OF MASS SPECTROMETRIC APPROACHES TO THE STRUCTURAL CHARACTERIZATION OF RECOMBINANT GLYCOPROTEINS

A strategy for the structural characterization of recombinant proteins that incorporates the mass spectrometric techniques described previously is presented in Fig. 4. This figure establishes a mass spectrometric scheme that relates sample issues (e.g., quantity of available protein, protein purity, degree of structural detail desired) with the time necessary to prepare and analyze the protein by each of the mass spectrometric techniques which have been discussed. With each of these mass

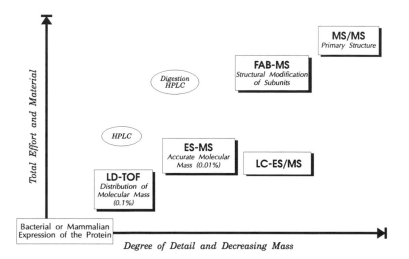

Figure 4 Mass spectrometric approaches for the structural characterization of recombinant proteins as a function of material, time, mass, and degree of structural detail. This graph illustrates the complementarity of various mass spectrometric techniques and permits the selection of a technique based on the level of structural detail. For instance, MALD-TOF can be used very early in a characterization strategy since it requires small amounts of material and can detect the molecular mass of intact proteins; however, the structural detail obtained is low, i.e., molecular mass. Tandem mass spectrometry requires more material and effort, but the level of detail is high, i.e., primary sequence. The circles indicate procedures which may be required prior to the next level of mass spectrometric analysis.

spectrometric techniques, the time of analysis is similar. What varies greatly is the type of information available by each technique as well as the time necessary to suitably prepare the sample for analysis. For example, the quantity of sample needed for MALD-TOF analysis is several picomoles and the quality of spectra are minimally influenced by various buffer components which may be present in a sample. This is in contrast to MS/MS where several hundred picomoles are needed for analysis and success is greatly influenced by the cleanliness (i.e., no involatile buffer components) of the sample. Additionally, while tandem mass spectrometry provides the greatest degree of structural detail, it also requires degradation of the protein and isolation of specific peptides of interest, which place additional time and material constraints on the analysis.

The strategy presented in Fig. 4 has been applied to the primary structural characterization of rhM-CSF. This example highlights the strengths and limitations of the various mass spectrometric approaches, as well as sample handling requirements which may be necessary prior to analysis.

rhM-CSF is a hematopoietic growth factor that induces proliferation of cells of the monocyte/macrophage lineage. The Chinese hamster ovary-expressed protein is a homodimer consisting of two 223 amino acid disulfide linked subunits which result from processing of a 553 amino acid precursor (Fig. 5). MALD-TOF mass spectrometric analysis of the intact glycoprotein reveals a molecular mass of ≈64,900 Da (Fig. 6). Nonreducing SDS–PAGE yields an apparent molecular mass of 90,000 Da. The large discrepancy between the SDS–PAGE-derived apparent molecular mass and the MALD-TOF molecular mass arises from the fact that the protein mobility in denaturing gel electro-

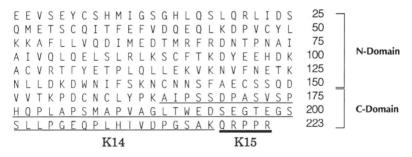

Figure 5 Primary sequence of CHO-derived rhM-CSF. The protein exists as a disulfide-linked 223 amino acid homodimer. When the protein is treated with a lysine- and/or arginine-specific protease under native conditions it is cleaved into three peptides, the N-terminal domain (1–163) and the C-terminal domain (164–223) which consists of two peptides, K-14 and K-15, as shown.

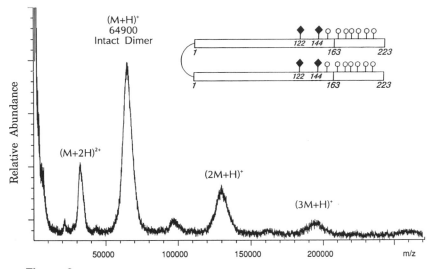

Figure 6 The matrix-assisted laser desorption time-of-flight mass spectrum of rhM-CSF. The observed molecular ion at ~64,900 corresponds to the average molecular mass for the glycosylated, disulfide-linked homodimer. A diagram depicting the basic structural features of this molecule is shown in the insert. The squares represent sites of *N*-linked glycosylation and the circles represent variable sites of *O*-linked glycosylation. In addition to the molecular ion, several other ions related to the sample are also observed. This is characteristic of MALD-TOF.

phoresis is not strictly a measure of the chemical mass. The gel mobility is influenced by both the amount of bound SDS as well as conformational features of the protein which cannot be predicted a priori. The MALD-TOF spectrum shown in Fig. 6 is characteristic of that for a glycoprotein. The native rhM-CSF homodimer is the predominant peak in the spectrum and exhibits an average mass of 64,900 Da. The spectrum is also characterized by peaks corresponding to doubly charged ions [e.g., $(M+2H)^{2+}$] as well as by oligomers of the main component [e.g., $(2M+H)^+$, $(3M+H)^+$]. One can quickly establish that the protein is post-translationally modified since the experimentally determined mass of the protein, 64,900 Da, is much larger than the predicted chemical mass of the dimeric protein, 50,253 Da. The broadness of the peak at 64,900 Da is reflective of the fact that MALD-TOF MS is a low resolution technique in comparison to other mass spectrometric approaches and that the sample consists of a number of discrete glycoforms. At this mass, the MALD-TOF MS is incapable of resolving the fine structure which results from the various glycoforms of the molecule. Nonetheless, MALD-TOF analysis can quickly provide some structural information, the benefit of which is very dependent upon the questions one seeks to answer. In this case, subsequent analyses, which are outlined below, have revealed that two classes of carbohydrate are present on the protein.

Combining MALD-TOF analysis with chemical, glycolytic, or enzymatic manipulation of the native protein provides additional information with which to interpret the MALD-TOF spectrum of the native molecule. Previous studies have shown that a C-terminal domain of native rhM-CSF is selectively cleaved with trypsin leaving the disulfide bonded N-terminal domain intact. Figure 7 shows the MALD-TOF spectrum of native rhM-CSF following trypsin treatment and reduction and alkylation. Similar to that shown previously, the spectrum is characterized by abundant molecular ions, doubly charged ions and oligomers. The peaks in the 24,000 to 26,000-Da mass range are less broad than in the native rhM-CSF spectrum, indicating that the C-terminal domain which was cleaved during trypsin treatment of the native dimer probably contains considerable carbohydrate heterogeneity. Furthermore, the N-terminal domain, which contains two consensus sequences for *N*-glycosylation, appears to be post-translationally modified since the MALD-TOF molecular mass is still approximately 5000 Da greater than that predicted for the alkylated N-terminal domain monomer. Interestingly, this mass spectrum appears to resolve two peaks at mass 24,639 and 25,040. This difference in mass corresponds to an *O*-linked glycan. However, from these data

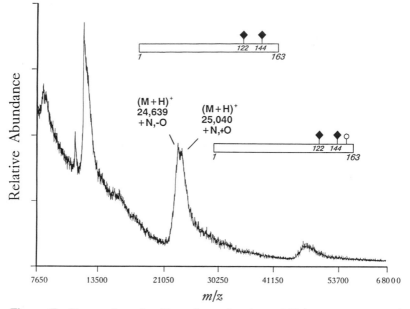

Figure 7   The matrix-assisted laser desorption time-of-flight mass spectrum of the molecular ion region of reduced and alkylated N-terminal domain of rhM-CSF. Two major peaks are observed which correspond to the N-terminal domain containing two sites of *N*-linked glycosylation *(m/z* = 24,639) and the N-terminal domain containing two *N*-linked and one *O*-linked site of glycosylation *(m/z* = 25,040). A diagram illustrating these structures is shown in the insert.

alone it would be difficult to attribute this difference to an O-glycan. Subsequent characterization has shown that the structures associated with these peaks are as shown in the figure insert. Each peak contains two *N*-linked carbohydrates and the peak at the higher mass does, as suspected, contain an additional site of *O*-linked carbohydrate.

In this example, these data provide overall structural information concerning the extent of glycosylation and the type of carbohydrate that may be present. Depending on the mass of the glycoprotein, the extent of sialylation, *N*-linked distribution, *O*-linked distribution, and the type of *N*-glycan (e.g., high mannose, hybrid or complex) may be established by comparison of MALD-TOF spectra prior to and subsequent to glycosidase digestion. Generally, these data do not pinpoint the location of these modifications. In the case of N-linked glycans, the Asn-X-Ser/Thr sequence motif is easily recognizable from the amino acid sequence and localizes the potential sites of N-glycans. However, while the presence or absence of glycan can be established, it is generally difficult to assign the site-specific location without further sample preparation and analysis.

The specific sites of N-linked glycosylation may be further delineated by proteolytic digestion of the protein followed by reverse-phase HPLC separation of the products and molecular mass analysis before and after treatment with a glycosidase. Figure 8 shows a typical ES mass spectrum from an achromobacter protease I peptide isolated from rhM-CSF and subsequently treated with PNGase F to remove the N-linked carbohydrate. ES mass spectra are characterized by a series of multiply charged ions, where each ion represents a different charge state of the ionized molecule. Since the mass spectrometric analysis is based on the mass-to-charge ratio, multiply charged ions of the same molecule appear in different regions of the mass spectrum, with the more highly charged species appearing at lower apparent mass. In this example, the spectrum shows charge states of the peptide ranging from $(M+6H)^{6+}$ at mass 551.5 to $(M+4H)^{4+}$ at mass 826.7. These ions can be deconvoluted using a mathematical transformation to generate a singly charged spectrum similar to those observed in FAB MS (insert Fig. 8). In positive ion spectra, the observed charge distribution is dependent on the number of accessible sites available for protonation. In the spectrum shown, the experimentally determined mass is 3303.2 and the theoretical mass is 3303.7, which represents an error of 0.015%.

As illustrated in Fig. 8, electrospray mass spectrometry offers an attractive alternative to FAB MS for obtaining accurate

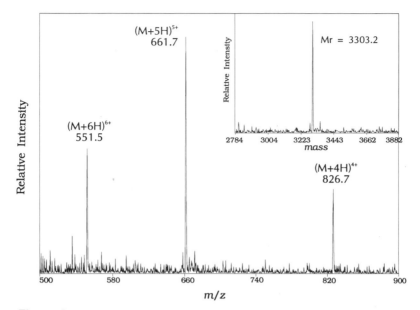

Figure 8 The electrospray mass spectrum of a PNGase F-treated achromobacter proteinase I peptide isolated from rhM-CSF. Three major peaks are observed in the mass spectrum corresponding to $(M+6H)^{6+}$, $(M+5H)^{5+}$, and $(M+4H)^{4+}$. Multiple peaks corresponding to the same molecule and differing only in the total number of charges are characteristic of ES MS. These data can be converted into a more familiar, mass versus intensity spectrum (see insert), in which only one peak is observed, in this example, corresponding to the average molecular weight of the peptide, 3303.2. This value is within 0.015% of the theoretical mass for this peptide.

molecular masses on both peptides and proteins. A major advantage of ES MS is its ability to directly couple with liquid chromatographic systems, thus making on-line analysis of proteins or proteolytic digests feasible. Furthermore, the technique appears to be more generally applicable to protein analysis than FAB MS and may provide higher resolution for protein analysis than MALD-TOF MS. However, since single components are represented in multiple charge states, mixtures may be difficult to deconvolute due to the complexity of the mass spectrum. For glycoprotcin analysis, ES MS does not seem to be as generally applicable as MALD-TOF MS.

As was noted previously, treatment of native rhM-CSF with trypsin or achromobacter protease I liberated a C-terminal domain that was thought to bc O-glycosylated based on MALD-TOF analysis. Reverse-phase high performance liquid chromatographic analysis of the products of this digestion revealed extensive heterogeneity in the K14 peptide of the C-terminal domain (Fig. 9). Upon neuraminidase and O-glycosidase digestion, the multiple peaks observed for the K14 peptide in Fig. 9 collapse to a single chromatographic peak which has the expected N-terminus of the K14 C-terminal domain peptide. In

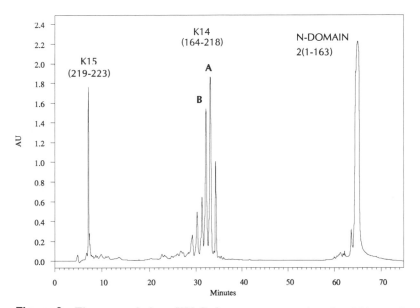

Figure 9 The reversed-phase HPLC chromatogram monitored at 214 nm of the products resulting from the digestion of the native rhM-CSF dimer with achromobacter proteinase I. The C-terminal domain consists of two peptides: K14 and K15. The K14 peak elutes as several peaks under these gradient conditions and each peak was collected and characterized. The FAB mass spectra of the molecular ion regions of the K14 peaks labeled A and B are shown in Fig. 10.

order to establish the site-specific nature of this heterogeneity, FAB MS was used to analyze the peaks as shown in Fig. 9. The FAB MS spectra of peaks A and B are shown in Figs. 10A and B. The masses of the major peaks in these spectra are consistent with the K14 peptide containing from two to four *O*-glycans with varying amounts of sialic acid. In most cases, the experimentally determined masses are within several tenths of a dalton from the theoretical values. Since the C-terminal domain contains many potential sites of *O*-glycosylation, the specific sites of glycosylation cannot be established from these data.

The spectra shown in Fig. 10 are typical of those obtained from FAB MS analysis, although these are of particularly high mass for glycopeptides. Generally, the spectra show a continuum of background due to chemical noise from the liquid matrix. Little, if any, sequence information is obtained if the analysis is performed on sample quantities of hundreds of picomoles or less. Higher quantities of sample often result in a limited amount of fragmentation. It is not always clear, however, whether these lower-molecular-weight ions result from fragmentation of the peptide of interest or from minor contaminants, and extreme caution must be used when interpreting peptide sequence from FAB MS spectra.

The techniques which have been discussed thus far primarily generate molecular weight information on peptides or proteins. In situations where molecular mass provides the structural information necessary for the solution to the problem, further discriminating analysis is not required. In the case of the *O*-glycosylated peptide cited previously, one might attempt to N-terminally sequence the peptide, and, based on the lack of expected signal at serine or threonine residues, conclude which are likely to be modified. However, it is not necessarily known if the absence of signal is due to glycosylation, phosphorylation, or other β-hydroxy amino acid modifications. In situations where there may be heterogeneity due to partial occupancy at a

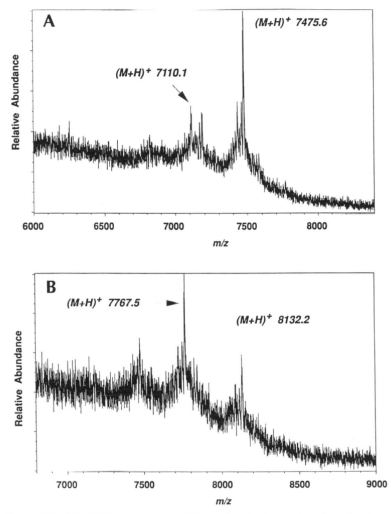

Figure 10 The FAB mass spectra of the molecular ion regions from fractions A and B of the K14 peptide isolated from the reverse-phase HPLC chromatogram of the native digest of rhM-CSF (Fig. 9). The protonated molecular ions correspond to the K14 peptide containing the following O-linked glycans: (A) 2(GalNAc-Gal) + 3NeuAc, $(M+H)^+$ 7110.1 and 3(GalNAc-Gal) + 3NeuAc, $(M+H)^+$ 7475.6; and (B) 3(GalNAc-Gal) + 4NeuAc, $(M+H)^+$ 7767.5 and 4(GalNAc-Gal) + 4NeuAc, $(M+H)^+$ 8132.2. The theoretical average molecular masses are 7110.4, 7475.8, 7767.0, and 8132.4, respectively, which are in excellent agreement with those determined experimentally.

single site, N-terminal sequence data may not provide an unambiguous interpretation. As shown by the FAB-MS data above, mass spectrometry does provide the opportunity to positively establish the chemical class of the post-translational modification and provides information about the heterogeneity within the peptide. To unequivocally assign the specific sites and type of modification within the protein, tandem mass spectrometry can be used *(145)*. Figure 11 shows the tandem mass spectrum from a peptide resulting from a neuraminidase-treated chymotryptic digestion of the C-terminal domain of rhM-CSF. The mass of the peptide at m/z 1967.9 establishes the location of the peptide within the C-terminal domain and that the peptide is modified by the addition of one site of O-linked glycosylation. Within this peptide there are four potential sites of O-linked carbohydrate attachment. The MS/MS spectrum reveals that the site of carbohydrate attachment occurs at the first of four closely spaced serines. This assignment is based primarily on the occurrence of the peak label w_{13} which signifies loss of the carbohydrate moiety from the indicated serine residue. There are also several peaks in the tandem mass spectrum that are related to the carbohydrate and these establish its sequence. MS/MS spectra are generally characterized by peaks at low mass, which are indicative of the amino acids present in the peptide, and high mass ions which result from R-group loss from the intact protein. Masses between the low and high mass ions result from sequence-specific fragmentation of the original peptide. These fragment ions can be assembled to establish the original linear sequence of amino acids within the peptide. Obviously, these spectra require considerable expertise for proper interpretation and may require substantial time for complete understanding. However, these data provide positive information on the nature of covalent modifications and provide the opportunity for obtaining sequence information on a peptide while in the presence of other peptides, minimizing the need for complete separation.

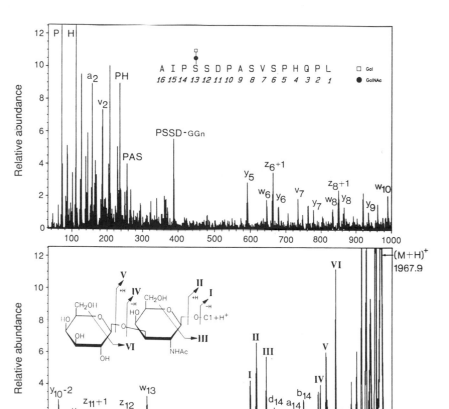

Figure 11 The tandem mass spectrum of a neuraminidase-treated, chymotryptic peptide isolated from the C-terminal domain of rhM-CSF. The protonated mass of 1967.9 corresponds to the peptide sequence shown containing one occupied site of *O*-linked glycosylation. The tandem mass spectrum verified the peptide sequence based on series of fragment ions from the C terminus (i.e., v,w,y, and z), identified the modification as a hexose-*N*-acetylhexosamine group (roman numerals), and the site of attachment as the first of the four closely spaced serines based on the w_{13} and y_{13} fragment ions.

151

CONCLUSIONS

Rapid changes in cellular and molecular biology now make it possible to achieve high-level heterologous protein expression in bacterial or mammalian hosts. Given these technological advances, the putative amino acid sequence can be inferred from the cDNA rather than through direct structural analysis of the expressed protein. Concurrent with these changes in cellular and molecular biology have been the development of new mass spectrometric techniques as well as the continued refinement of existing techniques. It has been the combination of these two events that has led to the widespread application of mass spectrometry to biological structural analysis. To date, the majority of biological mass spectrometry applications have been for the molecular mass analysis of peptides. However, recent advances in mass spectrometry have extended the useful mass range to greater than 200,000 Da, creating the possibility for mass spectrometric-based protein molecular mass analysis.

Given the variety of mass spectrometric methodologies, the choice of an appropriate technology is closely aligned with the question one seeks to answer. If the goal is to establish whether a protein expressed by two different cell types is processed similarly, a simple mass measurement on the intact protein may provide the necessary information. If, however, one wants to determine whether certain amino acids are post-translationally modified, molecular mass analysis of the intact protein alone would not provide the level of detail necessary to make this assessment. Proteolytic digestion and isolation of the peptide of interest followed by mass spectrometry and tandem mass spectrometry would most likely be required.

The quantity of available material must also be considered when designing a characterization strategy. Generally, the mass spectrometric methods providing the greatest structural detail require the largest quantity of sample and the most sample preparation. However, this must be balanced with the fact that mixture analysis and selective sequencing of one peptide in the presence of other peptides is much easier using mass spectrometry than using conventional approaches. The additional handling necessary to analyze a complex mixture by N-terminal sequence analysis may result in sample losses which exceed the total sample required for the mass spectrometry-based approach. Furthermore, the structural information obtained by conventional techniques may be less than that obtained by MS approaches due to the complexity of the protein.

New mass spectrometry developments and the refinement of existing technologies continue at a steady pace and the majority of these activities have direct applicability to biological analysis. What was difficult or impossible several years ago is, in many cases, routine today. The successful integration of established and new MS approaches for the structural characterization of proteins and glycoproteins requires both an understanding of the type of structural information that can be obtained as well as an appreciation of the biological problem.

Appendix

SEQUENCES OF COMMONLY USED PROTEASES

The sequences shown are the unprocessed forms of the active protease. The residue shown in **bold** marks the N-terminal residue of the processed protease.

Achromobacter Proteinase I

(EC 3.4.21.50) precursor— *Achromobacter lyticus*
Accession number: A32687

T. Ohara, K. Makino, H. Shinagawa, A. Nakata, S. Norika, and F. Sakiyama, *J. Biol. Chem.* **264**, 20,625 (1989)

```
         !         !         !         !         !
MKRICGSLLLLGLSISAALAAPASRPAAFDYANLSSVDKVALRTMPAVDV
AKAKAEDLQRDKRGDIPRFALAIDVDMTPQNSGAWFYTADGQFAVWRQRV
RSEKALSLNFGFTDYYMPAGGRLLVYPATQAPAGDRGLISQYDASNNNSA
RQLWTAVVPGAEAVIEAVIPRDKVGEFKLRLTKVNHDYVGFGPLARRLAA
ASGEKGVSGSCNIDVVCPEGDGRRDIIRAVGAYSKSGTLACTGSLVNNTA
NDRKMYFLTAHHCGMGTASTAASIVVYWNYQNSTCRAPNTPASGANGDGS
MSQTQSGSTVKATYATSDFTLLELNNAANPAFNLFWAGWDRRDQNYPGAI
AIHHPNVAEKRISNSTSPTSFVAWGGGAGTTHLNVQWQPSGGVTEPGSSG
SPIYSPEKRVLGQLHGGPSSCSATGTNRSDQYGRVFTSWTGGGAAASRLS
DWLDPASTGAQFIDGLDSGGGTPNTPPVANFTSTTSGLTATFTDSSTDSD
GSIASRSWNFGDGSTSTATNPSKTYAAAGTYTVTLTVTDNGGATNTKTGS
VTVSGGPGAQTYTNDTDVAIPDNATVESPITVSGRTGNGSATTPIQVTIY
HTYKSDLKVDLVAPDGTVYNLHNRTGGSAHNIIQTFTKDLSSEAAQRAPG
SCG
```

A Practical Guide to Protein and Peptide Purification for Microsequencing, Second Edition

Appendix

Bovine Pancreatic Chymotrypsinogen

Chymotrypsin (EC 3.4.21.1) B precursor — Bovine
Accession number: A00953

L.B. Smillie, A. Furka, N. Nagabhushan, K.J. Stevenson, and
C.O. Parkes, *Nature* (London) *218*, 343 (1968)

```
.........!.........!.........!.........!.........!
CGVPAIQPVLSGLARIVNGEDAVPGSWPWQVSLQDSTGFHFCGGSLISED
WVVTAAHCGVTTSDVVVAGEFDQGLETEDTQVLKIGKVFKNPKFSILTVR
NDITLLKLATPAQFSETVSAVCLPSADEDFPAGMLCATTGWGKTKYNALK
TPDKLQQATLPIVSNTDCRKYWGSRVTDVMICAGASGVSSCMGDSGGPLV
CQKNGAWTLAGIVSWGSSTCSTSTPAVYARVTALMPWVQETLAAN
```

Bovine Pancreatic Trypsin

Trypsinogen (EC 3.4.21.4) — Bovine
Accession number: A00946

K. Titani, L.H. Ericsson, H. Neurath, and K.A. Walsh, *Biochemistry 14*, 1358 (1975)

```
.........!.........!.........!.........!.........!
VDDDDKIVGGYTCGANTVPYQVSLNSGYHFCGGSLINSQWVVSAAHCYKS
GIQVRLGEDNINVVEGNEQFISASKSIVHPSYNSNTLNNDIMLIKLKSAA
SLNSRVASISLPTSCASAGTQCLISGWGNTKSSGTSYPDVLKCLKAPILS
DSSCKSAYPGQITSNMFCAGYLEGGKDSCQGDSGGPVVCSGKLQGIVSWG
SGCAQKNKPGVYTKVCNYVSWIKQTIASN
```

Papain

Papain (EC 3.4.22.2) — Papaya
Accession number: A00974

R.E.J. Mitchel, I.M. Chaiken, and E.L. Smith, *J. Biol. Chem.*
245 3485 (1970)

```
.........!.........!.........!.........!.........!
IPEYVDWRQKGAVTPVKNQGSCGSCWAFSAVVTIEGIIKIRTGNLNQYSE
QELLDCDRRSYGCNGGYPWSALQLVAQYGIHYRNTPYYEGVQRYCRSREK
GPYAAKTDGVRQVQPYNQGALLYSIANQPVSVVLQAAGKDFQLYRGGIFV
GPCGNKVDHAVAAVGYNPGYILIKNSWGTGWGENGYIRIKRGTGNSYGVC
GLYTSSFYPVKN
```

References

Elution of Blotted Proteins from Nitrocellulose and
  PVDF Membranes
Amino Acid Analysis of Electroblotted Samples
Internal Sequence from Electroblotted and
  Electroeluted Samples
Partial Fragmentation of Proteins in SDS-Gel Slices
Removal of SDS from Protein Samples
Microscale Transfer of Proteins to Volatile Buffers
Mass Spectrometric Identification of Peptide
  Sequences and Post-translational Modifications
N-Terminal Modifications *in Vivo*
Computer Analysis of Protein Sequences

GENERAL REFERENCES FOR PROTEIN/PEPTIDE PURIFICATION AND MICROSEQUENCING

1. Methods in Enzymology "Enzyme Structure Part I" (C.H.W. Hirs and S.N. Timasheff, eds.) Vol. 91, Academic Press, New York, 1983.

2. Methods of Protein Microcharacterization (J.E. Shively, ed.), Humana Press, Clifton, New Jersey, 1986.

3. Methods in Protein Sequence Analysis (K.A. Walsh, ed.), Humana Press, Clifton, New Jersey, 1987.

4. Macromolecular Sequencing and Synthesis: Selected Methods and Applications (D.H. Schlesinger, ed.), R. Liss, Inc., New York, 1988.

5. Protein/Peptide Sequence Analysis: Current Methodologies (A.S. Bhown, ed.), CRC Press, Boca Raton, Florida, 1988.

6. Sequence of Proteins and Peptides (G. Allen, ed.), Elsevier Science, Amsterdam, 1989.

7. Protein Sequencing: A Practical Approach (J.B.C. Findlay and M.J. Geisow, eds.), IRL Press, Oxford, 1989

8. Techniques in Protein Chemistry I (J.J. Villafranca, ed.), Academic Press, San Diego, 1992.

9. Methods in Enzymology "Guide to Protein Purification" (M.P. Deutscher, ed.), Vol. 182, Academic Press, New York, 1990.

10. Techniques in Protein Chemistry II (J.J. Villafranca, ed.), Academic Press, San Diego, 1991.

GENERAL DESCRIPTION OF MODERN APPROACHES TO PROTEIN MICROSEQUENCING

11. M. Hunkapiller, S. Kent, M. Caruthers, W. Dreyer, J. Firca, C. Giffin, S. Horvath, T. Hunkapiller, P. Tempst, and L. Hood, "A Microchemical Facility for the Analysis and Synthesis of Genes and Proteins," *Nature* **310**, 105 (1984).

12. M.W. Hunkapiller, J.E. Strickler, and K.J. Wilson, "Contemporary Methodology for Protein Structure Determination," *Science* **226**, 304 (1984).

13. S. Kent, L. Hood, R. Aebersold, D. Teplow, L. Smith, V. Farnsworth, P. Cartier, W. Hines, P. Hughes, and C. Dodd, "Approaches to Sub-Picomole Protein Sequencing," *Bio-Techniques* **5**, 314 (1987).

14. R., Aebersold, G.D. Pipes, R.E. Wettenhall, H. Nika, and L.E. Hood, "Covalent Attachment of Peptides for High Sensitivity Solid-Phase Sequence Analysis." *Anal. Biochem.* **187**, 56 (1990).

15. J.M. Coull, D.J.C. Pappin, J. Mark, R. Aebersold, and H. Koester "Functionalized Membrane Supports for Covalent Protein Microsequence Analysis," *Anal. Biochem* **194**, 110 (1991).

GENERAL OVERVIEW OF AUTOMATED GAS-PHASE SEQUENCING

16. M.W. Hunkapiller, R.M. Hewick, W.J. Dreyer, and L.E. Hood, "High-Sensitivity Sequencing with a Gas-Phase Sequenator," *Methods Enzymol.* 91, 399 (1983).

17. M.W. Hunkapiller, K. Granlund-Moyer, and N.W. Whitcley, "Gas-Phase Protein/Peptide Sequencer" *in* Methods of Protein Microcharacterization (J.E. Shively, ed.), Humana Press, Clifton, New Jersey, 1986.

18. V. Farnsworth, W. Carson, and H. Krutzsch, "Reduced Chemical Background Noise in Automated Protein Sequencers," *Peptide Res.* **4**, 245 (1991).

MANUAL GAS-PHASE SEQUENCING METHODS

19. G.E. Tarr, "Manual Edman Sequencing" *in* Methods of Protein Microcharacterization (J.E. Shively, ed.), p. 155, Humana Press, Clifton, New Jersey, 1986.

20. W.F. Brandt and G. Frank, "Manual Gas-Phase Isothiocyanate Degradation," *Anal. Biochem.* **168**, 314 (1988)

21. M. Haniu and J.E. Shively, "Microsequence Analysis of Peptides and Proteins. IX. Manual Gas-Phase Microsequencing of Multiple Samples," *Anal. Biochem.* **173**, 296 (1988).

AMINO ACID ANALYSIS

22. M.W. Hunkapiller, K. Grundlund-Moyer, and N.W. Whiteley, "Analysis of Phenylthiohydantoin Amino Acids by HPLC" *in* Methods of Protein Microcharacterization (J.E. Shively, ed.), p. 315, Humana Press, Clifton, New Jersey, 1986.

23. R. Knecht and J.-Y. Chang, "High Sensitivity Amino Acid Analysis Using DABS-Cl Precolumn Derivatization Method" *in* Advanced Methods in Protein Microsequence Analysis (B. Wittmann-Liebold, J. Salnikow, and V.A. Erdmann, eds.), Springer-Verlag, Berlin, 1986.

24. B.A. Bidlingmeyer, T.L. Tarvin, and S.A. Cohen, "Amino Acid Analysis of Submicrogram Hydrolyzate Samples" *in* Methods in Protein Sequence Analysis (K.E. Walsh, ed.), p. 229, Humana Press, Clifton, New Jersey, 1986.

25. V. Stocchi, G. Piccoli, M. Magnani, F. Palma, B. Biagiarelli, and L. Cucchiarini, "Reversed-Phase High-Performance Liquid Chromatography Separation of Dimethylaminoazobenzene Sulfonyl- and Dimethylaminoazobenzene Thiohydantoin-Amino Acid Derivatives for Amino Acid Analysis and Microsequencing Studies at the Picomole Level," *Anal. Biochem.* **178**, 107 (1988).

CHEMICAL MODIFICATION METHODS

26. A.M. Crestfield, S. Moore, and W.H. Stein "The Preparation and Enzymatic Hydrolysis of Reduced and S-Carboxymethylated Proteins," *J. Biol. Chem.* **238**, 622 (1963).

27. A.N. Glazer, R.J. Delange, and D.S. Sigman. "Chemical Modification of Protein" *in* Laboratory Techniques in Biochemistry and Molecular Biology (T.S. Work and E. Work, eds.), North-Holland/American Elsevier, Amsterdam, 1975.

28. P.C. Andrews and J.E. Dixon, "A Procedure for *in Situ* Alkylation of Cystine Residues on Glass Fiber prior to Protein Microsequence Analysis," *Anal. Biochem.* **161**, 524 (1987).

29. R.L. Lundblad, Chemical Reagents for Protein Modification, CRC, Press Boca Raton, Florida, 1991.

PROTEOLYTIC AND CHEMICAL CLEAVAGE METHODS

30. E. Gross, "The Cyanogen Bromide Reaction," *Methods Enzymol.* 11, 238 (1967).

31. W.A. Schroeder, J.B. Shelton, and J.R. Shelton, "An Examination of Conditions for the Cleavage of Polypeptide Chains with Cyanogen Bromide: Application to Catalase," *Arch. Biochem. Biophys.* **130**, 551 (1969).

32. C. B. Kaspar, "Fragmentation of Proteins for Sequence Studies and Separation of Peptide Mixtures" *in* Protein Sequence Determination (Needleman, S.B., ed.), pp. 114-161, Springer-Verlag, New York, 1970.

33. A. Fontana, "Modification of Tryptophan with BNPS-Skatole (2-(2-nitrophenylsulfenyl)-3-methyl-3-bromoindolenine)," *Methods. Enzymol.* 25, 419 (1972).

34. J. Houmard and G.R. Drapeau, "Staphylococcal Protease: A Proteolytic Enzyme Specific for Glutamoyl Bonds," *Proc. Natl. Acad. Sci. USA* **69**, 3506 (1972).

35. M. Levy, L. Fishman, and I. Shenkein, "Mouse Submaxillary Gland Proteases," *Methods Enzymol.* 19, 672 (1972).

36. W.E. Brown and F. Wold, "Alkyl Isocyanates as Active-Site Specific Reagents for Serine Proteases. Reaction Properties," *Biochemistry* **12**, 828 (1973).

37. I. Schechter, A. Patchornike, and Y. Burstein, "Selective Chemical Cleavage of Tryptophenyl Peptide Bonds by Oxidative Chlorination with N-Chlorosuccimimide," *Biochemistry* **15**, 5071 (1976).

38. D. N. Podell and G.N. Abraham, "A Technique for the Removal of Pyrogluatmic Acid from the Amino Terminus of Proteins Using Calf Liver Pyroglutamate Aminopeptidase," *Biochem. Biophys. Res. Commun.* **81**, 176 (1978).

39. W.C. Mahoney and M.A. Hermodson, "High Yield Cleavage of Tryptophenyl Peptide Bonds by o-Iodosobenzoic Acid," *Biochemistry* **18**, 3810 (1979).

40. G.R. Drapeau, "Substrate Specificity of a Proteolytic Enzyme Isolated from a Mutant of *Pseudomonas fragi*," *J. Biol. Chem.* **255**, 839 (1980).

41. T. Masaki, M. Tanabe, K. Nakamura, and M. Soejima, "Studies on a New Proteolytic Enzyme from *Achromobacter lyticus* M497-1. I. Purification and Some Enzymatic Properties," *Biochim. Biophys. Acta* **660**, 44 (1981).

42. M.A. Hermodson, "Chemical Cleavage of Proteins" *in* Methods in Protein Sequence Analysis, (M. Elzinga ed.), p. 313, Humana Press, Clifton, New Jersey, 1982.

43. A. Fontana, D. Dalzoppo, C. Grandi, and M. Zambonin, "Cleavage at Tryptophan with o-Iodosobenzoic Acid," *Methods Enzymol.* 91, 311 (1983).

44. P. Jekel, W. Weijer, and J. Beintema, "Use of Endoproteinase Lys-C from *Lysobacter enzymogenes* in Protein Sequence Analysis," *Anal. Biochem.* **134**, 347 (1983).

45. N. C. Price and C. M. Johnson, "Proteinases as Probes of Conformation of Soluble Proteins" *in* Proteolytic Enzymes: A Practical Approach (R.J. Beynon and J.S. Bond, eds.), p. 163, IRL Press, Oxford, 1989.

46. S. Ishi, Y. Abe, H. Matsushita, and I. Kato, "An Asparaginyl Endopeptidase Purified from Jack Bean Seeds," *J. Protein Chem.* **9**, 294 (1990).

47. M. Moyer, A. Harper, G. Payne, J. Ryals, and D. Fowler, "*In Situ* Digestion with Pyroglutamate Aminopeptidase for N-Terminal Sequencing Electroblotted Proteins," *J. Protein Chem.* **9**, 282 (1990)

48. L. R. Riviere, M. Fleming, C. Elicone, and P. Tempst, "Study and Applications of the Effects of Detergents and Chaotropes on Enzymatic Proteolysis" *in* Techniques in Protein Chemistry II (J. J. Villafranca, ed.), p. 171, Academic Press, New York, 1991.

49. K. L. Stone, D. E. McNulty, M. L. LoPresti, J. M. Crawford, R. DeAngelis, and K. R. Williams, "Election and Internal Amino Acid Sequencing of PVDF-Blotted Proteins" *in* Techniques in Protein Chemistry III (R. Angeletti, ed.), Academic Press, San Diego, 1992.

HPLC PEPTIDE ISOLATION

50. R.J. Simpson and E.D. Nice, "The Role of Microbore HPLC in the Purification of Subnanomole Amounts of Polypeptides and Proteins for Gas-Phase Sequence Analysis" *in* Methods in Protein Sequence Analysis, (K.E. Walsh, ed.), p. 213, Humana Press, Clifton, New Jersey, 1986.

51. J.E. Shively, "Reverse Phase HPLC Isolation and Microsequence Analysis" *in* Methods of Protein Microcharacterization (J.E. Shively, ed.), Humana Press, Clifton, New Jersey, 1986.

52. K. L. Stone, and K. R. Williams, "High Performance Liquid Chromatographic Peptide Mapping and Amino Acid Analysis in the Sub-Nanomole Range," *J. Chrom.* **359**, 203 (1986).

53. K.R. Williams, K.L. Stone, M.K. Fritz, B.M. Merrill, W.H. Konigsberg, M. Pandolfo, O. Valentini, S. Riva, W. Reddigari, G.L. Patel, and J.W. Chase, "Use of HPLC Comparative Peptide Mapping in Structure/Function Studies" *in* "Proteins: Structure and Function" (J.J. L'Italien, ed.), p. 45, Plenum Press, 1987.

54. K.L. Stone and K.R. Williams, "Small Bore HPLC Purification of Peptides in the Sub-nanomole Range" *in* Macromolecular Sequencing and Synthesis, (D. Schlesinger, ed.), p. 7, A. R. Liss, 1988.

55. R.J. Simpson, R.L. Moritz, G.S. Begg, M.R. Rubira, and E.C. Nice, "Micropreparative Procedures for High-Sensitivity Sequencing of Peptides and Proteins," *Anal. Biochem.* **177**, 221 (1989).

56. K.L. Stone, M.B. LoPresti, N.D. Williams, J.M. Crawford, R. DeAngelis, and K.R. Williams, "Enzymatic Digestion of Proteins and HPLC Peptide Isolation in the Sub-nanomole Range" *in* Techniques in Protein Chemistry (T. Hugli, ed.), p. 377, Academic Press, 1989.

57. K.L. Stone, and J.I. Elliott, G. Peterson, W. McMurray, and K.R. Williams, "Reversed Phase HPLC Fractionation of Enzymatic Digests and Chemical Cleavage Products of Proteins" *in* Methods in Enzymology (J. McCloskey, ed.), Vol 193, p. 389, Academic Press, New York, 1990.

58. K.L. Stone, M.B. LoPresti, and K.R. Williams, "Enzymatic Digestion of Proteins and HPLC Peptide Isolation in the Sub-Nanomole Range" *in* Laboratory Methodology in Biochemistry (C. Fini, A. Floridi, and V. Finelli, eds.) p. 181, CRC Press, Boca Raton, Florida, 1990.

59. L.D. Ward, G.E. Rcid, R.L. Moritz, and R.J. Simpson, "Peptide Mapping and Internal Sequencing of Proteins from Acrylamide Gels" *in* Current Research in Protein Chemistry: Techniques, Structure, and Function (J. J. Villafranca, ed.), pp. 179–190, Academic Press, New York, 1990.

60. K.L. Stone, M.B. LoPresti, J.M. Crawford, R. DeAngelis, and K.R. Williams, "Reverse-Phase HPLC Separation of Sub-nanomole Amounts of Peptides Obtained from Enzymatic Digests" *in* HPLC of Peptides and Proteins: Separation, Analysis and Conformation (C. Mant and R.S. Hodges, eds.), pp. 669–677, CRC Press, Boca Raton, Florda, 1991.

SDS–PAGE ELECTROPHORESIS/ ELECTROBLOTTING METHODS

61. Gel Electrophoresis of Proteins (B.D. Hammes and and D. Rickwood, eds.), IRL Press, Oxford, 1981.

62. U.K. Laemmli, "Cleavage of Structural Proteins during the Assembly of the Head of Bacteriophage T4," *Nature (London)* **227**, 680 (1970)

63. W. Schaffner and C. Weissman, "A Rapid, Sensitive, and Specific Method for the Determination of Protein in Dilute Solution," *Anal. Biochem.* **56**, 502 (1973).

64. P.H. O'Farrell, "High Resolution Two-Dimensional Electrophoresis of Proteins," *J. Biol. Chem.* **250**, 4007 (1975).

65. H. Towbin, T. Staehelin, and J. Gordon, "Electrophoretic Transfer of Proteins from Polyacrylamide Gels to Nitrocellulose Sheets: Procedure and Some Applications," *Proc. Natl. Acad. Sci. USA* **76**, 4350 (1979).

66. C. Schafer-Nielsen, P. J. Svendsen, and C. Rose, "Separation of Macromolecules in Isotachophoresis Systems Involving Single or Multiple Counterions," *J. Biochem, Biophys Methods* **3**, 97 (1980).

67. J.M. Gershoni and G.E. Palade, "Protein Blotting: Principles and Applications," *Anal. Biochem.* **131**, 1 (1983).

68. M.G. Pluskal, M.B. Przekop, M.R. Kavonian, C. Vecoli, and D.A. Hicks, "Immobilon PVDF Transfer Membrane: A New Membrane Substrate for Western Blotting of Proteins," *Biotechniques* **4**, 272 (1986).

69. M. Moermans, G. Daneels, M. DeRaeymaeker, B. DeWever, and J. DeMey "The Use of Colloidal Metal Particles in Protein Blotting," *Electrophoresis*, **8**, 403 (1987).

70. H. Schagger and G. von Jagow, "Tricine-Sodium Dodecyl Sulfate-Polyacrylamide Gel Electrophoresis for Separation of Proteins in the Range from 1 to 100 kDa," *Anal. Biochem.* **166**, 368 (1987).

71. H. Gultekin and K. Heermann, "The Use of Polyvinyldene-Difluoride Membranes as a General Blotting Matrix," *Anal. Biochem.* **172**, 320 (1988).

72. J. Reig and D.C. Klein, "Submicrogram Quantities of Unstained Proteins Are Visualized on Polyvinylidene Difluoride Membranes by Transillumination," *Applied and Theoretical Electrophoresis* **1**, 59 (1988).

73. J. Hughes, K. Mack, and V. Hamparian, "India Ink Staining of Proteins on Nylon and Hydrophobic Membranes," *Anal. Biochem.* **173**, 18 (1988).

74. J. Kyhse-Andersen, "Electroblotting of Multiple Gels: A Simple Apparatus without Buffer Tank for Rapid Transfer of Proteins from Polyacrylamide to Nitrocellulose," *J. Biochem. Biophys. Methods* **10**, 203 (1984).

75. J. M. Coull and D. J. C. Pappin, "A Rapid Fluorescent Staining Procedure for Proteins Electroblotted onto PVDF Membranes," *J. Protein Chem.* **9**, 259 (1990).

RECOVERY OF PROTEINS FROM SDS GELS: ELECTROELUTION AND DIFFUSION

76. A.S. Bhown and J.C. Bennett, "High-Sensitivity Sequence Analysis of Proteins Recovered from Sodium Dodecyl Sulfate Gels," *Methods Enzmol.* **91**, 450 (1983).

77. M.W. Hunkapiller, E. Lujan, F. Ostrander, and L.E. Hood, "Isolation of Microgram Quantities of Proteins from Polyacrylamide Gels for Amino Acid Sequence Analysis," *Methods Enzymol.* **91**, 227 (1983).

78. J.J. L'Italien, "Solid Phase Methods in Protein Sequence Analysis" *in* "Methods of Protein Microcharacterization" (J.E. Shively, ed.), p. 298, Humana Press, Clifton, New Jersey, 1986.

79. M.W. Hunkapiller and E. Lujan "Purification of Microgram Quantitites of Proteins by Polyacrylamide Gel Electrophoresis" *in* Methods of Protein Microcharacterization (J.E. Shively, ed.), p. 89, Humana Press, Clifton, New Jersey, 1986.

RECOVERY OF PROTEINS FROM SDS GELS: ELECTROBLOTTING ONTO PVDF (IMMOBILON) MEMBRANES AND TREATED GLASS FIBER FILTERS

80. J. Vandekerckhove, G. Bauw, M. Puype, J. Van Damme, and M. Van Montagu, "Protein-Blotting on Polybrene-Coated Glass-Fiber Sheets. A Bases of Acid Hydrolysis and Gas-Phase Sequencing of Picomole Quantities of Protein Previously Separated on SDS-Polyacrylamide Gel," *Eur. J. Biochem.* **152**, 9 (1985).

81. R.H. Aebersold, D.B. Teplow, L.E. Hood, and S.B. Kent, "Electroblotting onto Activated Glass: High Efficiency Preparation of Protiens from Analytical Sodium Dodecyl Sulfate-Polyacrylamide Gels for Direct Sequence Analysis," *J. Biol. Chem.* **261**, 4229 (1986).

82. J. Vandekerckhove, G. Bauw, J. Van Damme, M. Puype, and M. Van Montagu, "Protein-blotting from SDS-Polyacrylamide Gels on Glass-Fiber Sheets Coated with Quaternized Ammonium Polybases" *in* Methods in Protein Sequence Analysis (K.A. Walsh, ed.), p. 261, Humana Press, Clifton, New Jersey, 1986.

83. G. Bauw, M. De Loose, D. Inze, M. Van Montagu, and J. Vandekerchove, "Alterations in the Phenotype of Plant Cells Studied by Amino-Terminal Amino Acid-Sequence Analysis of Proteins Electroblotted from Two-Dimensional Gel-Separated Total Extracts," *Proc. Natl. Acad. Sci. USA* **84**, 4806 (1987).

84. P. Matsudaira, "Sequence from Picomole Quantities of Proteins Electroblotted onto Polyvinylidene Difluoride Membranes," *J. Biol. Chem.* **262**, 10035 (1987).

85. C. Eckerskorn, W. Mewes, H. Goretzki, and F. Lottspeich, "A New Siliconized-Glass Fiber as Support for Protein-Chemical Analysis of Electroblotted Proteins," *Eur. J. Biochem.* **176**, 509 (1988).

86. N. LeGendre and P. Matsudaira, "Direct Protein Microsequencing from Immobilon-P Transfer Membrane," *BioTechniques* **6**, 154 (1988).

87. M. Moos, Jr., N.Y. Nguyen, and T.Y. Liu "Reproducible High Yield Sequencing of Proteins Electrophoretically Separated and Transferred to an Inert Support," *J. Biol. Chem.* **263**, 6005 (1988).

88. Q-Y. Xu and J.E. Shively, "Microsequence Analysis of Peptides and Proteins. VIII. Improved Electroblotting of Proteins onto Membranes and Derivatized Glass-Fiber Sheets," *Anal. Biochem.* **170**, 19 (1988).

89. S.W. Yuen, A.H. Chui, K.J. Wilson, and P.M. Yuan, "Microanalysis of SDS-PAGE Electroblotted Proteins," *BioTechniques* **7**, 74, (1989).

90. M. Mansfield and A. Weiss, "Factors Affecting Protein Retention and 'Blow-through' in Protein Blotting Applications," FASEB, 75th Annual Meeting, Atlanta, Georgia, 1991.

ELECTROBLOTTING OF PROTEINS FROM ISOELECTRIC FOCUSING GELS

91. J.-C. Hsieh, F.P. Lin, and M.F. Tam, "Electroblotting onto Glass-Filter from an Analytical Isoelectrofocusing Gel: A Preparative Method for Isolating Proteins for N-Terminal Microsequencing," *Anal. Biochem.* **170**, 1 (1988).

92. M. Knierm, J. Buchholz, and W. Pflug, "Electrotransfer of Proteins after Isoelectric Focusing (with or without Urea) on Fabric-Reinforced, Immobilized pH Gradient Gels," *Anal. Biochem.* **172**, 139 (1988).

93. M. Ogata, K. Suzuki, and Y. Satoh, "Characterization of Human Residual Catalase of an Acatalasemic Patient by Isoelectric-Focusing and Sodium Dodecyl Sulfate-Polyacrylamide Gel Electrophoresis followed by Electrophoretic Blotting and Immunodetection," *Electrophoresis* **10**, 194 (1989).

ELUTION OF BLOTTED PROTEINS FROM NITROCELLULOSE AND PVDF MEMBRANES

94. O. Salinovich and R.C. Montelaro, "Reversible Staining and Peptide Mapping of Proteins Transferred to Nitrocellulose after Separation by Sodium Dodecylsulfate-Polyacrylamide Gel Electrophoresis," *Anal. Biochem.* **156**, 341 (1986).

95. B. Szewczyk and D. Summers, "Preparative Elution of Proteins Blotted to Immobilon Membranes," *Anal. Biochem.* **168**, 48 (1988).

96. D. L. Crimmins, D. W. McCourt, R. S. Thoma, M. G. Scott, K. Macke, and B. D. Schwartz, "*In Situ* Chemical Cleavage of Proteins Immobilized to Glass-Fiber and Polyvinylidenedifluoride Membranes: Cleavage at Tryptophan Residues with (2-(2′-Nitrophenylsulfonyl)-3-methyl-3′-bromoindolenine) to Obtain Internal Amino Acid Sequence," *Anal. Biochem.* **187**, 27 (1990).

AMINO ACID ANALYSIS OF ELECTROBLOTTED SAMPLES

97. E. Hildebrandt and V.A. Fried, "Phosphoamino Acid Analysis of Protein Immobilized on Polyvinylidene Difluoride Membrane," *Anal. Biochem.* **177**, 407 (1989).

98. M.P. Kamps and B.M. Sefton, "Acid and Base Hydrolysis of Phosphoproteins Bound to Immobilon Facilitates Analysis of Phosphoamino Acids in Gel-Fractionated Proteins," *Anal. Biochem.* **176**, 22 (1989).

99. G.I. Tous, J.L. Fausnaugh, O. Akinyosoye, H. Lackland, P. Winter-Cash, F.J. Victoria, and S. Stein, "Amino Acid Analysis on Polyvinylidene Difluoride Membranes," *Anal. Biochem.* **179**, 50 (1989).

INTERNAL SEQUENCE FROM ELECTROBLOTTED AND ELECTROELUTED SAMPLES

100. R.H. Aebersold, J. Leavitt, R.A. Saavedra, L.E. Hood, and S.B. Kent, "Internal Amino Acid Sequence Analysis of Proteins Separated by One- or Two-dimensional Gel Electrophoresis after *in Situ* Protease Digestion on Nitrocellulose," *Proc. Natl. Acad. Sci. USA* **84**, 6970 (1987).

101. M.G. Scott, D.L. Crimmins, D.W. McCourt, J.J. Tarrand, M.C. Eyerman, and M.H. Nahm, "A Simple *in Situ* Cyanogen Bromide Cleavage Method to Obtain Internal Amino Acid Sequence of Proteins Electroblotted to Polyvinylidifluoride Membranes," *Biochem. Biophys. Res. Commun.* **155**, 1353 (1988).

102. T.E. Kennedy, M.A. Gawinowicz, A. Barzilai, E.R. Kandel, and J.D. Sweatt, "Sequencing of Proteins from Two-dimensional Gels by *in Situ* Digestion and Transfer of Peptides to Polyvinylidene Difluoride Membranes: Application to Proteins Associated with Sensitization in Aplysisa," *Proc. Natl. Acad. Sci. USA* **85**, 7008 (1988).

103. T.E. Kennedy, K.Wager-Smith, A. Barzilai, E.R. Kandel, and J.D. Sweatt, "Sequencing Proteins from Acrylamide Gels," *Nature (London)* **336**, 499 (1988).

104. P. Tempst, A. J. Link, L. R. Riviere, M. Fleming, and C. Elicone, "Internal Sequence of Proteins Separated on Polyacrylamide Gels at the Submicrogram Level: Improved Methods, Applications and Gene Cloning Strategies," *Electrophoresis* **11**, 537 (1990).

105. J. Fernandez, M. DeMott, D. Atherton, and S. M. Mische, "Internal Protein Sequence Analysis: Enzymatic Digestion for Less Than 10 μg of Protein Bound to Polyvinylidene Difluoride or Nitrocellulose Membranes," *Anal. Bioc.* **201**, 255-264 (1992)

106. K.L. Stone, D.E. McNulty, M.L. LoPresti, J.M. Crawford, R. DeAngelis, and K.R. Williams, "Elution and Internal Amino Acid Sequencing of PVDF-Blotted Proteins" *in* Techniques in Protein Chemistry III (R. Angeletti, ed.), pp. 23–34, Academic Press, San Diego, 1992.

PARTIAL FRAGMENTATION OF PROTEINS IN SDS-GEL SLICES

107. D.W. Cleveland, S.G. Fischer, M.W. Kirschner, and U.K. Laemmli, "Peptide Mapping by Limited Proteolysis in So-

dium Dodecyl Sulfate and Analysis by Gel Electrophoresis," *J. Biol. Chem.* **252**, 1102 (1977).

108. V. Nikodem and J.R. Fresco, "Protein Fingerprinting by SDS-Gel Electrophoresis after Partial Fragmentation with CNBr," *Anal. Biochem.* **97**, 382 (1979).

109. S. Detke and J.M. Keller, "Comparison of the Proteins Present in HeLa Cell Interphase Nucleoskeletons and Metaphase Chromosome Scaffolds," *J. Biol. Chem.* **257**, 3905 (1982).

110. C.J.M. Saris, J. Van Eenbergen, B.G. Jenks, and H.P.J. Bloemers, "Hydroxylamine Cleavage of Proteins in Polyacrylamide Gels," *Anal. Biochem.* **132**, 54 (1983).

111. I. Walker and C.W. Anderson, "Partial Proteolytic Protein Maps: Cleveland Revisited," *Anal. Biochem.* **146**, 108 (1985).

112. S. Mahboub, C. Richard, A. Delacourte, and K.-K. Han, "Applications of Chemical Cleavage Procedures to the Peptide Mapping of Neurofilament Triplet Protein Bands in Sodium Dodecyl Sulfate-Polyacrylamide Gel Electrophoresis," *Anal. Biochem.* **154**, 171 (1986).

113. H. Kawasaki and K. Suzuki, "Separation of Peptides Dissolved in a Sodium Dodecyl Sulfate Solution by Reversed-Phase Liquid Chromatography: Removal of Sodium Dodecyl Sulfate from Peptides Using an Ion-Exchange Precolumn" *Anal. Biochem.* **186**. 264 (1990).

113a. J. Rosenfield, J. Capdevielle, J. Guillemot, and P. Ferrara, "In Gel Digestions of Proteins for Internal Sequence Analysis after 1 or 2 Dimensional Gel Electrophoresis," *Anal. Biochem.* **203**, 173 (1992)

REMOVAL OF SDS FROM PROTEIN SAMPLES

114. D. A. Hager and R. R. Burgess, "Elution of Proteins from Sodium Dodecyl Sulfate-Polyacrylamide Gels, Removal of Sodium Dodecyl Sulfate, and Renaturation of Enzymatic Activity: Results with Sigma Subunit of *Eschericia coli* RNA Polymerase, Wheat Germ DNA Topoisomerase, and Other Enzymes," *Anal. Biochem.* **109**, 76 (1980).

115. S.N. Vinogradov and O.H. Kapp, "Removal of Sodium Dodecyl-Sulfate from Proteins by Ion-Retardation Chromatography," *Methods Enzymol.* **91**, 259 (1983).

116. W.H. Konigsberg and L. Henderson, "Removal of Sodium Dodecyl-Sulfate from Proteins by Ion-Pair Extraction," *Methods Enzymol.* **91**, 254 (1983).

117. H. Suzuki and T. Terada, "Removal of Dodecyl Sulfate from Protein Solution," *Anal. Biochem.* **172**, 259 (1988).

118. H. Kawasaki and K. Suzuki, "Separation of Peptides Dissolved in a Sodium Dodecyl Sulfate Solution by Reversed-Phase Liquid Chromatography: Removal of Sodium Dodecyl Sulfate from Peptides Using an Ion-Exchange Precolumn," *Anal. Biochem.* **186**, 264 (1990).

119. H. Kawasaki, Y. Emori, and K. Suzuki, "Production and Separation of Peptides from Proteins Stained with Coomassie Brilliant Blue R-250 after Separation by Sodium Dodecyl Sulfate-Polyacrylamide Gel Electrophoresis," *Anal. Biochem.* **191**, 332 (1990).

MICROSCALE TRANSFER OF PROTEINS TO VOLATILE BUFFERS

120. R.J. Simpson, R.L. Moritz, E.E. Nice, and B. Grego, "A High-Performance Liquid Chromatography Procedure for Recovering Subnanomole Amounts of Protein from SDS-Gel Electroeluates for Gas-Phase Sequence Analysis," *Eur. J. Biochem.* **165**, 21 (1987).

MASS SPECTROMETRIC IDENTIFICATION OF PEPTIDE SEQUENCES AND POST-TRANSLATIONAL MODIFICATIONS

121. M. Barber, R.S. Bordoli, G.J. Elliott, R.D. Sedgwick, and A.N. Tyler, "Fast Atom Bombardment Mass Spectrometry," *Anal. Chem.* **54**, 645A (1982).

122. K. Biemann, "Mass Spectrometry of Peptides and Proteins," *Annu. Rev. Biochem.* **61**, 977 (1992).

123. H.A. Scoble and S.A. Martin, "Characterization of Recombinant Proteins," *Methods in Enzymol.* **193**, 519 (1990).

124. S.A. Carr, M.E. Hemling, M.F. Bean, and G.D. Roberts, "Integration of Mass Spectrometry in Analytical Biotechnology," *Anal. Chem.* **63**, 2802 (1991).

125. K. Biemann and H.A. Scoble, "Characterization by Tandem Mass Spectrometry of Structural Modifications in Proteins," *Science* **239**, 992 (1987).

126. D.F. Hunt, J.R. Yates III, J. Shabanowitz, S. Winston, and C.R. Hauer, "Protein Sequencing by Tandem Mass Spectrometry," *Proc. Natl. Acad. Sci. USA* **83**, 6233 (1986).

127. F. Hillenkamp, M. Karas, R.C. Beavis, and B.T. Chait, "Matrix-Assisted Laser Desorption/Ionization Mass Spectrometry of Biopolymers," *Anal. Chem.* **63**, 1193A (1991).

128. M. Mann, "Electrospray: Its Potential and Limitations as an Ionization Method for Biomolecules," *Organic Mass Spectrom.* **25**, 575 (1990).

129. M. Karas, D. Bachmann, U. Bahr, and F. Hillenkamp, "Matrix-Assisted Ultraviolet-Laser Desorption of Non-volatile Compounds," *Int. J. Mass Spectrom. Ion Proc.* **78**, 53 (1987).

130. J.A. Hill, R.S. Annan, and K. Biemann, "Matrix-Assisted Laser Desorption Ionization with a Magnetic Mass Spectrometer," *Rapid Commun. Mass Spectrom.* **5**, 395 (1991).

131. R.L. Hettich and M.V. Buchanan, "Investigation of UV Matrix-Assisted Laser Desorption Fourier Transform Mass Spectrometry for Peptides," *J. Am. Soc. Mass Spectrom.* **2**, 22 (1991).

132. R.C. Beavis and B.T. Chait, "Rapid, Sensitive Analysis of Protein Mixtures by Mass Spectrometry," *Proc. Natl. Acad. Sci. USA* **87**, 6873 (1990).

133. R.C. Beavis and B.T. Chait, "Factors Affecting the Ultraviolet Laser Desorption of Proteins," *Rapid Commun. Mass Spectrom.* **3**, 233 (1989).

134. M.M. Siegel, I.J. Hollander, P.R. Hamann, J.P. James, L. Hinman, B.J. Smith, A.P.H. Farnsworth, A. Phipps, D.J. King, M. Karas, A. Ingendoh, and F. Hillenkamp, "Matrix-

Assisted UV-Laser Desorption/Ionization Mass Spectrometric Analysis of Monoclonal Antibodies for the Determination of Carbohydrate, Conjugated Chelator, and Conjugated Drug Content," *Anal. Chem.* **63**, 2470 (1991).

135. H. Egge, J. Peter-Katalinic, M. Karas, and B. Stahl, "The Use of Fast Atom Bombardment and Laser Desorption Mass Spectrometry in The Analysis of Complex Carbohydrates," *Pure Appl. Chem.* **63**, 491 (1991).

136. T. Huth-Fehre, J.N. Gosine, K.J. Wu, and C.H. Becker, "Matrix-Assisted Laser Desorption Mass Spectrometry of Oligodeoxythymidylic Acids," *Rapid Commun. Mass Spectrom.* **6**, 209 (1992).

137. P.R. Griffin, J.A. Coffman, L.E. Hood, and J.R. Yates III, "Structural Analysis of Proteins by Capillary HPLC Electrospray Tandem Mass Spectrometry," *Intl. J. Mass Spectrom. Ion Proc.* **111**, 131 (1991).

138. D.F. Hunt, R.A. Henderson, J. Shabanowitz, K. Sakaguchi, H. Michel, N. Sevilir, A.L. Cox, E. Appella, and V.H. Engelhard, "Characterization of Peptides Bound to the Class I MHC Molecule HLA-A2.1 by Mass Spectrometry," *Science* **255**, 1261 (1992).

139. W. Aberth, K.M. Straub, and A.L. Burlingame, "Secondary Ion Mass Spectrometry with Cesium Ion Primary Beam and Liquid Target Matrix for Analysis of Bioorganic Compounds," *Anal. Chem.* **54**, 2029 (1982).

140. M. Barber, R.S. Bordoli, R.D. Sedgwick, and A.N. Tyler, "Fast Atom Bombardment of Solids (F.A.B.): A New Ion Source for Mass Spectrometry," *J.C.S. Chem. Comm.* **7**, 325 (1981).

141. K. Biemann, "Amino Acid Sequence in Oligopeptides and Proteins," *in* Biochemical Applications of Mass Spectrometry (G.R. Waller and O.C. Dermer, eds.) p. 469, Wiley, New York, 1980.

142. E. De Pauw, "Liquid Matrices for Secondary Ion Mass Spectrometry," *Mass Spectrom. Rev.* **5**, 191 (1986).

143. R.S. Johnson, S.A. Martin, K. Biemann, J.T. Stults, and J.T. Watson, "Novel Fragmentation Process of Peptides by Collision-Induced Decomposition in a Tandem Mass Spectrometer: Differentiation of Leucine and Isoleucine," *Anal. Chem.* **59**, 2621 (1987).

144. R.S. Johnson, S.A. Martin, and K. Biemann, "Collision-Induced Fragmentation of $(M+H)^+$ Ions of Peptides: Side Chain Specific Sequence Ions," *Intl J. Mass Spectrom. Ion Proc.* **86**, 137 (1988).

145. C.A. Settineri, K.F. Medzihradszky, F.R. Masiarz, A.L. Burlingame, C. Chu, and C. George-Nascimento, "Characterization of O-Glycosylation Sites in Recombinant B-Chain of Platelet-derived Growth Factor Expressed in Yeast Using Liquid Secondary Ion Mass Spectrometry, Tandem Mass Spectrometry and Edman Sequence Analysis," *Biomed. Environ. Mass Spectrom.* **19**, 665 (1990).

N-TERMINAL MODIFICATIONS *IN VIVO*

146. J.L. Brown and W.K. Roberts, "Evidence that Approximately Eighty Percent of the Soluble Proteins from Ehrlich Ascites Cells Are Amino-Terminally Acetylated," *J. Biol. Chem.* **251**, 1009 (1976).

147. J.L. Brown, "A Comparison of the Turnover of Amino-Terminally Acetylated and Nonacetylated Mouse L-Cell Proteins," *J. Biol. Chem.* **254**, 1447 (1979).

148. H.P. Driessen, W.W. De Jong, G.I. Tesser, and H. Bloemendal, Critical Reviews in Biochemistry (G.D. Fasman, ed.), Vol. 18, pp. 281-325, CRC Press, Boca Raton, Florida, 1985.

149. D. Wellner, C. Panneerselvam, and B. L. Horecker, "Sequencing of Peptides and Proteins with Blocked N-terminal Amino Acids: N-Acetylserine or N-Acetylthreonine," *Proc. Natl. Acad. Sci. USA* **87**, 1947 (1990).

150. H. Hirano, S. Komatsu, H. Takakura, F. Sakiyama, and S. Tsunasawa, "Deblocking and Subsequent Microsequence Analysis of N^α-Blocked Proteins Electroblotted onto PVDF Membrane," J. Biochem. 111, 754 (1992).

151. M. Mitta, K. Asada, Y. Uchimura, F. Kimizuka, I. Kato, F. Sakiyama, and S. Tsunasawa, "The Primary Structure of Porcine Liver Acylamino Acid Releasing Enzyme Deduced from cDNA Sequence," *J. Biochem.* **106**; 548 (1989)

COMPUTER ANALYSIS OF PROTEIN SEQUENCES

152. G.R. Reeck, C. de Haen, D.C. Teller, R. F. Doolittle, W.M. Fitch, R. E. Dickerson, P. Chambon, A. D. McLachlan, E. Margoliash, T.H. Jukes, and E. Zuckerkandl, " "Homology" in Proteins and Nucleic Acids: A Terminology Muddle and a Way out of It," *Cell* **50**, 667 (1987).

153. Methods in Enzymology "Molecular Evolution: Computer Analysis of Protein and Nucleic Acid Sequences" (R. F. Doolittle, ed.), Vol. 183, Academic Press, New York, 1990.

154. G. von Heijne, Sequence Analysis in Molecular Biology, Academic Press, New York, 1988.

155. R. F. Doolittle, Of URFS and ORFS: A Primer on How to Analyze Derived Amino Acid Sequences, Univ. Sci. Books, Mill Valley, California, 1986.

156. P. Argos, "Computer Analysis of Protein Structure," *Methods Enzymol.* 182, 751 (1990)

157. M. O. Dayhoff, W. C. Barker, and L. T. Hunt, "Establishing Homologies in Protein Sequences" *Methods Enzymol.* **91**, 525 (1983).

158. M. Gribskov and J. Devereux (eds.) Sequence Analysis Primer, Stockton Press, New York, 1991.

159. W. R. Pearson and D. J. Lipman, "Improved Tools for Biological Sequence Comparison," *Proc. Natl. Acad. Sci. USA* **85**, 2444 (1988).

160. P. M. McCaldon and P. Argos, "Oligopeptide Biases in Protein Sequences and Their Use in Predicting Proteins Coding Regions in Nucleotide Sequences," *Proteins* **4**, 99 (1988).

161. J. F. Harper, M. R. Sussman, G. E. Schaller, C. Putnam-Evans, H. Charbonneau, and A. C. Harmon, "A Calcium-Dependent Protein Kinase with a Regulatory Domain Similar to Calmodulin," *Science* **252** (1991).

162. T. P. Hopp and K. R. Woods, "Prediction of Protein Antigenic Determinants from Amino Acid Sequences," *Proc. Natl. Acad. Sci. USA* **78**, 3824 (1981).

163. J. Kyte and R. F. Doolittle, "A Simple Method for Displaying the Hydropathic Character of a Protein," *J. Mol. Biol.* **157**, 105 (1982).